# Applied
# Animal Reproduction

# Applied
# Animal Reproduction

H. Joe Bearden
and
John W. Fuquay
*Mississippi State University*

Reston Publishing Company, Inc.
A Prentice-Hall Company
Reston, Virginia

**Library of Congress Cataloging in Publication Data**

Bearden, Henry Joe
  Applied animal reproduction.  .

  Includes bibliographies and index.
  1. Veterinary physiology.   2. Reproduction.
3. Veterinary obstetrics.   4. Livestock—
Breeding.   I. Fuquay, John, joint author.
II. Title
SF768.B42        636.08′2        79-28155
ISBN 0-8359-0249-8

Drawings by George H. Taylor
Photographs by Marco W. Nicovich

©1980 by
Reston Publishing Company, Inc.
A Prentice-Hall Company
Reston, Virginia

10   9   8   7   6   5   4   3   2   1

Printed in the United States of America

# Table of contents

Table 1-1.  Accuracy of sire proofs for predicting milk yield, 3
Table 2-1.  Reproductive organs of the female with their major functions, 9
Table 3-1.  Reproductive organs of the male with their major functions, 23
Table 4-1.  Molecular size of peptide and protein hormones which regulate production, 38
Table 4-2.  Hormones that regulate reproduction, 40
Table 4-3.  Major steroid hormones produced by the gonads, 44
Table 5-1.  Species differences in various characteristics of the estrous cycle, 54
Table 5-2.  Species and breed differences in age and weight at puberty, 54
Table 5-3.  Primary characteristics of the periods of the estrous cycle in the cow, 55
Table 6-1.  Effect of time of insemination on ovulation and fertility in cows, 74
Table 7-1.  Transport time of ova in the oviduct of farm animals, 79
Table 7-2.  Estimated fertile life of sperm and ova in farm animals, 83
Table 7-3.  Effect of age of the ovum on fertility in cattle, 83
Table 7-4.  Effect of the length of time of storage of extended semen on its fertility level and the difference between 1-mo. and 5-mo. nonreturns, 83
Table 8-1.  Species and breed differences in gestation length, 86
Table 8-2.  Time comparisons from ovulation during cleavage for different farm species, 88
Table 8-3.  Certain organs that have been identified as forming from specific germ layers, 89
Table 8-4.  Developmental features in the cow and pig during differentiation, 93
Table 8-5.  Weight changes of the bovine uterus and its contents during pregnancy, 95

# Preface

This text is intended to give the undergraduate student majoring in animal or dairy science a complete overview of the reproductive processes. It is assumed that these students will have a limited background in physiology. Therefore, a major effort has been made to maintain clarity. It is hoped that this style of writing will also encourage use of this text in two-year agricultural curricula and in short courses where participants have a more limited educational background. Thirty years' experience in teaching a practical course in physiology of reproduction to students with a wide divergence of backgrounds has influenced the level of writing and the organization of the book. Comments and suggestions from students were given careful consideration during the preparation of the text.

Parts 1 and 2 are designed to help students develop both the terminology needed to discuss problems associated with physiology of reproduction and an understanding of the physiological processes controlling reproduction. These sections are up-to-date and include recent information on gonadal proteins and regulation of hormone receptor sites. Chapter 4 will be difficult because the concept of endocrine regulation will be new to most undergraduate students. When this information is reinforced in later chapters on reproductive processes in the female and the male, these concepts will seem less troublesome. Early introduction permits development of a more profound understanding of the endocrine regulation of reproduction.

Parts 3, 4, and 5 emphasize the application of basic concepts to the management of reproduction in livestock. This text is unique in the emphasis that is given to the applied aspects of reproduction. Five chapters are devoted to artificial insemination. These include collection, evaluation, storage, and utilization of semen through artificial insemination. Five chapters are written on reproductive management with specific chapters on environmental management, nutritional management, pregnancy diagnosis, and diseases affecting reproduction. The goal of these chapters goes beyond description of simple techniques for good reproductive management. They are designed to help students understand the rationale and principles used in developing guidelines for good reproductive management.

Several steps have been taken to make this text more readable. Important terms are italicized and defined when first introduced. Only the most prevalent theories are presented and these have been simplified rather than presented in lengthy discussions on the pros and cons of these concepts. A consensus is presented where disagreement exists in the literature. Also, reference citations are not listed in the text. A carefully selected reading list is included at the end of each chapter for those who wish to pursue specific topics in more depth. The references have been limited to encourage additional reading, rather than overwhelming students with the vast number of publications available on each topic. The writing style used in this book may be troublesome to instructors who are accustomed to delving into scientific literature. A number of referenced texts are currently available while a few have been written specifically for undergraduates. The selected references after each chapter include both reviews and publications of original research which will be useful for documentation.

H. Joe Bearden
John W. Fuquay

# Chapter

# 1

# Introduction and history

Reproduction is a complex science, so much so that Dr. William Hansel of Cornell University told his classes, "It is not a wonder that reproduction sometimes fails, but rather a miracle that so many pregnancies terminate successfully." This is obviously an overstatement. However, it does indicate the complexity of the subject. In order to understand the science of reproduction, it is necessary to include anatomy, physiology, endocrinology, embryology, histology, cytology, microbiology and some nutrition. Reproduction involves a series of physiological and psychological events that must be properly timed. The endocrine system, through the production of several hormones, is responsible for this timing.

Much of the existing knowledge pertaining to reproduction has been generated during the past 25 years. However, Aristotle (384–322 B.C.) wrote the first scientific papers on embryology. While most of his work was surprisingly accurate, his greatest contribution was to turn man's mind away from superstition, replacing it with observation. His great work was so far ahead of the age in which he lived it was nearly 2,000 years before anything of significance was added. Working without a microscope, Aristotle had to speculate on some things and here he fell into error. He formulated two theories for embryonic development: (1) the embryo was preformed and grew or enlarged during development; or (2) it was the result of differentiation from a formless being. He decided on the latter, crediting the popular belief that slime and decaying matter were capable of producing living animals. He described the human embryo as being organized out of the mother's menstrual blood. New developments awaited the invention of the microscope.

In 1668 Redi published papers based on his observations that tended to disprove Aristotle's theory that the human embryo developed from menstrual blood. de Graaf described the ovarian follicle in 1672, while Hamm and Leewenhook, with the aid of the early microscope, observed and described the human spermatozoa in 1677. During the period from 1759 to 1769 Wolf observed parts of the chick embryo take shape and virtually destroyed the preformation theory. However, he was not able to show that development began with a single cell. He set forth the theory of *Epigenesis*—development of an embryo by progressive growth and differentiation. Some individuals tended to hang onto the preformation theory until about 1860. Some believed that the preformed animal existed in the ovum and that growth was stimulated by seminal fluid from the male. Others believed that the preformed animal existed in the sperm cell which grew after entering the egg. Pasteur in 1864 demonstrated that bacteria reproduced themselves through cell division, thus destroying the theory of spontaneous generation. Dreish in 1900 provided the final proof that life came from single cells by separating daughter cells of a fertilized egg and showed that they could each develop into an embryo.

Reproduction has at least three purposes:

**1.** *Perpetuation of the species.* The strongest desire of an individual in any species including man is to maintain itself. The strongest impulse in an individual is saving its own life. Reproduction is nature's second strongest impulse, thus, the maintenance of the species.

**2.** *To provide food.* Man has learned to manage both domestic and wild species so that surplus animals may be harvested to supply meat. Through selection he has developed the milk-producing capabilities of a few species, so that milk, too, has become an important link in the human food chain.

**3.** *Genetic improvement.* The management and alteration of natural reproductive processes have been utilized as genetic tools.

Genetic improvement of any species is accomplished by selecting males and females with superior transmitting ability as parents of succeeding generations. Selection pressures have been applied for thousands of years, but more genetic progress has been made during the past 25 years than in any previous 100-year period. Some of this progress has been made through genetic knowledge and techniques; however, a large part of it has been accomplished by manipulating, or harnessing, the reproductive process.

The rate of genetic improvement through selection depends on several factors. *Variation*—that difference in production level for individual animals for characteristics such as milk production, rate of gain, weaning weight, etc., must exist. Otherwise there would be no basis for selection. *Heritability*—the percentage of total variation that is controlled by the genetic make up of the individual. This portion of the variation is also referred to as *genetic variation*. If all of the difference in the records made by different individuals was due to genetic variation, heritability would be one or 100%; thus an animal's production records or appearance would perfectly measure its genetic value. Heritability for most economic traits in farm animals ranges from 0 to 60%; milk production 25%; number of pigs per litter 5 to 10%; yearling body weight in sheep 20 to 59%; feed lot gain in beef cattle 50 to 55% and fertility in cattle 5%. *Environmental variation*—the difference between total variation and genetic variation. In order to measure genetic gain it is necessary to either control or account for environmental variation. *Selection intensity*—a factor that affects the rate of genetic improvement by determining the ratio of offspring utilized for extensive breeding versus the number culled after adequate sampling. The higher the culling rate, the faster genetic progress should be. *Generation interval*—the average interval between birth of an animal and the birth of its offspring. The shorter the generation interval, the faster progress can be made. Selection intensity can be changed at will, but only minor changes can be made in generation interval. From these accurately measured factors one can estimate the rate of genetic progress per generation or per year.

Artificial insemination is an example where reproduction processes have been used as a genetic tool. The true transmitting ability for milk production and overall type score in dairy bulls was virtually impossible to obtain prior to the introduction of artificial insemination (AI). The wide variation that exists, plus a relatively low heritability of the traits, requires a large number of offspring with production records or type scores. These records also need to be made in many different herds to account for environmental variation. Table 1-1 shows the accuracy of sire proofs for predicting milk yields, assuming heritability of 25% and an environmental correlation of 6% among daughters in the same herd. An *environmental correlation* exists when daughters of a sire are treated more alike than unrelated cows. Note that

TABLE 1-1. Accuracy of sire proofs for predicting milk yield

| Number of daughters | Accuracy when all cows are in different herds | Accuracy when all cows are in one herd |
|---|---|---|
| | % | % |
| 1 | 25 | 25 |
| 5 | 50 | 46 |
| 10 | 63 | 53 |
| 20 | 76 | 60 |
| 40 | 85 | 65 |
| 50 | 88 | 66 |
| 70 | 91 | 67 |
| 100 | 93 | 68 |
| 200 | 96 | 69 |
| 1000 | 99 | 70 |

From Schmidt and Van Vleck, *Principles of Dairy Science*, W.H. Freeman and Co., copyright © 1974.

when all daughters are in different herds the accuracy reaches 99% only when there are 1,000 daughters with records. When all daughters are in one herd the accuracy is only 70% with records on 1,000 daughters. Before AI, a bull seldom had more than 20 daughters and usually these were in one or possibly two herds. The example given here is for milk yield, but the general principle applies to essentially all economic traits of all species of farm animals. Traits with higher heritability will require fewer numbers. For traits with extremely low heritability, such as fertility in cattle, little progress can be expected through selection.

Other reproductive processes that have been or may be used as genetic tools are:

1. Frozen semen
2. Separation of male and female producing sperm
3. Synchronization of estrus
4. Superovulation
5. Ovum transfer
6. Storing fertilized ova
7. Test tube fertilization
8. Environmental influence on puberty

These processes will be discussed in detail in later chapters.

A tremendous amount of knowledge pertaining to reproduction has been generated since 1900, but the subject remains very dynamic. Reproductive efficiency has been greatly enhanced as improved practices based on new scientific information have been put into use. Greater use of these practices by those who use this book will result in further improvements.

Recent research has developed new hormone assay and hormone receptor site techniques that are constantly providing new information. Some of these techniques will no doubt rank with the discovery of the microscope in helping us to understand the reproductive processes. The student must keep in mind that the concepts presented in this book are as current as our present knowledge will allow but that future research will change some of them. It will be only through continued basic research that major improvements in reproductive efficiency will be accomplished.

**Suggested reading**

Asdell, S. A. 1977. Historical Introduction. *Reproduction in Domestic Animals*. (3rd ed.) Ed. H. H. Cole and P. T. Cupps. Academic Press.

Salisbury, G. W., N. L. VanDemark and J. R. Lodge. 1978. *Physiology of Reproduction and Artificial Insemination of Cattle*. (2nd ed.). W. H. Freeman and Co.

Schmidt, G. H. and L. D. Van Vleck. 1974. *Principles of Dairy Science*. W. H. Freeman and Co.

# Anatomy, function, and regulation

# Chapter

# 2

# The female reproductive system

The female reproductive system, as illustrated for the cow in Figure 2-1, consists of two ovaries and the female duct system. The duct system includes the oviducts, uterus, cervix, vagina, and vulva. The embryonic origin of the ovary is the *secondary sex cords of the genital ridge*. The genital ridges are first seen in the embryo as a slight thickening near the kidneys. The duct system originates from the *mullerian ducts*, a pair of ducts which appear during early embryonic development (see Chapter 8). An overview of the organs of reproduction for the female and the major functions of these organs are shown in Table 2-1.

**2-1**
**Ovaries**

The *ovaries* are considered the primary reproductive organs in the female. They are primary because they produce the female gamete (the *ovum*) and the female sex hormones (estrogens and progestins). The cow, mare, and ewe are *monotoccus*, normally giving birth to one young each *gestation period*. Therefore, one ovum is produced each estrous cycle. The sow is *polytoccus*, producing 10 to 25 ova each estrous cycle and giving birth to several young each gestation period.

The ovary of the cow is described as almond shaped but the shape is altered by growing follicles or corpora lutea. The average size is about 35 × 25 × 15 mm. The size will vary among cows, and active ovaries are larger than inactive ovaries. Therefore, one ovary is frequently larger than the other in a given cow. The ovaries of the ewe are almond shaped and less than half the size of those of a cow. The ovaries in the mare are kidney shaped and are two to three times larger than ovaries of a cow. The ovaries in the sow are slightly larger than those found in the ewe and appear as a "cluster of grapes" because of the extensive follicle growth and associated corpora lutea.

The ovary is composed of the *medulla* and its outer shell, the *cortex*. The medulla is composed primarily of blood vessels, nerves, and connective tissue. The cortex contains those cell and tissue layers associated with ovum

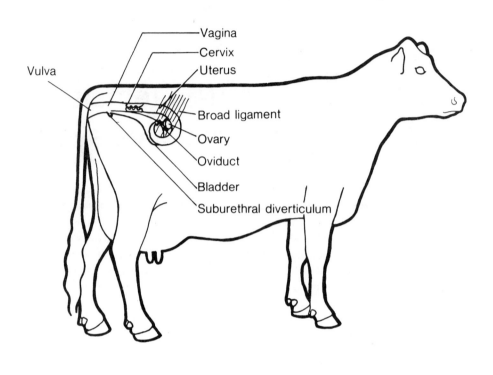

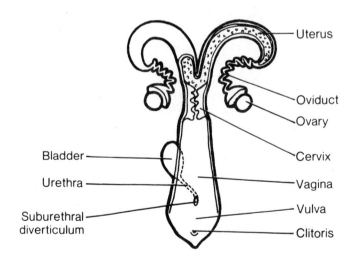

FIGURE 2-1 Reproductive system and associated parts of the urinary system of the cow as it appears in the natural state (top) and excised (bottom).

TABLE 2-1.   Reproductive organs of the female with their major functions

| Organ | Function(s) |
| --- | --- |
| Ovary | Production of oocytes<br>Production of estrogens (Graafian follicle)<br>Production of progestins (corpus luteum) |
| Oviduct | Gamete transport (spermatozoa and oocytes)<br>Site of fertilization |
| Uterus | Retains and nourishes the embryo and fetus |
| Cervix | Prevents microbial contamination of uterus<br>Reservoir for semen and transport of spermatozoa<br>Site of semen deposit during natural mating in sows and mares |
| Vagina | Organ of copulation<br>Site of semen deposit during natural mating in cows and ewes<br>Birth canal |
| Vulva | External opening to reproductive tract |

and hormone production. The outermost layer of the cortex of the ovary is the *surface epithelium*. The surface epithelium, a single layer of cuboidal cells, was originally called the germinal epithelium because it was believed to be the origin of female germ cells (oogonia). It is now known that germ cells do not arise from this epithelial layer. They arise from embryonic gut tissue and then migrate to the cortex of the embryonic gonad. Just beneath the surface epithelium is a thin, dense layer of connective tissue, the *tunica albuginea ovarii*. Below the tunica albuginea ovarii is the *parenchyma*, known as the functional layer because it contains ovarian follicles and the cells which produce ovarian hormones.

It is accepted that all primary follicles are formed during the prenatal period of the female. The greatest number are found in the fetal gilt 50 to 90 days postconception; the fetal calf 110 to 130 days postconception. A *primary follicle* is a germ cell surrounded by a single layer of follicular cells. They are located in the parenchyma and are frequently seen in groups called *egg nests*. It is estimated that approximately 75,000 primary follicles are found in the ovaries of a young calf. With continual follicular growth and maturation throughout her reproductive life, an old cow may have only 2500 potential ova. Some potential ova reach full maturity and are released into the duct system for possible fertilization and development of offspring. Most start development and become *atretic* (i.e., they degenerate). Therefore, the potential harvest of ova that could produce offspring is far greater than is actually realized.

Follicles are in a constant state of growth and maturation. A histological section of the cortex of a reproductively active female will reveal these maturation stages (Figure 2-2). The primary follicle has been described. This stage is followed by a proliferation of follicular cells surrounding the potential ovum. A potential ovum surrounded by two or more layers of follicular cells is a *secondary follicle*. Later in the development, an *antrum* (cavity) will form by fluid collecting between the follicular cells and separating them. When the antrum has formed, the follicle is classified as a *tertiary follicle*. The mature tertiary follicle, which appears as a fluid-filled blister on the surface of the ovary, is also called a *Graafian follicle*. The fluid in the antrum of a tertiary follicle is rich in estrogens and is called *liquor folliculi*.

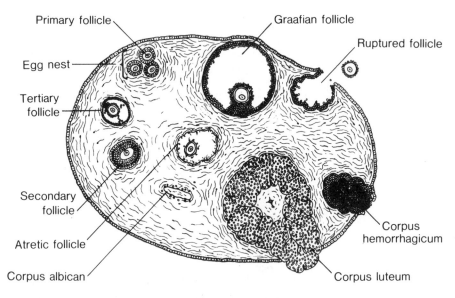

FIGURE 2-2. Diagram of structures that can be identified in a cross-section of an ovary of a reproductively active female. Different maturation stages for follicles and the corpus luteum can be observed. (Adapted from Patten. 1964. *Foundations of Embryology*. (2nd ed.) McGraw-Hill.)

Several cell layers in the Graafian follicle have been identified and are of functional importance (Figure 2-3). The outer, more fibrous cell layer is the *theca externa*. Inside this layer is the *theca interna*. These two cell layers are supplied with blood by a capillary network and can be differentiated microscopically with special histological staining techniques. A basement membrane separates the theca interna from the innermost cell layer, the *granulosa cells*, and prevents entry of the vascular system into these cells (Figure 2-4). The granulosa cells surround the antrum. In addition, the

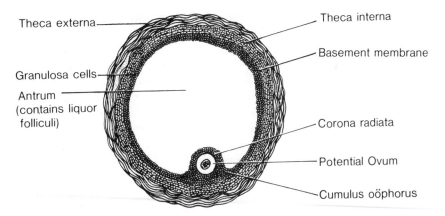

FIGURE 2-3.   Functionally important features of a Graafian follicle. (Redrawn from Hafez. 1974. *Reproduction in Farm Animals.* (3rd ed.) Lea and Febiger.)

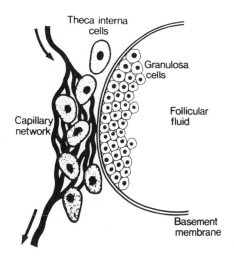

FIGURE 2-4.   Structure of the wall of the Graafian follicle showing how the granulosa cells are deprived of a blood supply by the basement membrane. (Austin and Short. 1972. *Reproduction in Mammals. 3. Hormones of Reproduction.* Cambridge University Press.)

*cumulus oophorus*, a hillock (mound) of granulosa cells, is located at one side of the antrum. The potential ovum rests upon the cumulus oophorus with other granulosa cells extending around the potential ovum. The granulosa cells surrounding and in immediate contact with the potential ovum are termed the *corona radiata*. Both theca interna and granulosa cells may be involved in the production of estrogen. One accepted theory states that the theca interna cells produce testosterone, which diffuses through the basement membrane for conversion to estrogen by the granulosa cells. Granulosa cells are the progesterone producing cells in the corpus luteum. When ovulation occurs, the follicle ruptures expelling the liquor folliculi, some granulosa cells, and the potential ovum (oocyte) into the body cavity near the opening to the oviduct. At the time of expulsion, the oocyte is surrounded by the corona radiata and a sticky mass containing other granulosa (*cumulus*) cells which aid the oviduct in picking up the ooctye and moving it down the oviduct. In some species the corona radiata is present at the time of fertilization. In other species these cells are shed quickly and are not present when fertilization occurs.

With rupture of the follicle, bleeding occurs and a blood clot forms at the ovulation site. The ruptured follicle with its blood filled cavity is called a *corpus hemorrhagicum*. The corpus hemorrhagicum is replaced by the *corpus luteum*, which forms rapidly by a proliferation of the theca externa, theca interna, and granulosa cells. Granulosa cells form the main component of the corpus luteum. The corpus luteum is a solid, yellowish body which produces progesterone and other progestins. The corpus luteum is the only ovarian source of these progestins. In a Holstein heifer, between 1 and 4 days of the cycle the average diameter of the corpus luteum is 8 mm. Between 5 and 9 days, it has grown to an average of 15 mm. The average maximum size of 20.5 mm is reached in 15 or 16 days in a heifer that is not pregnant, then it regresses in size with an average diameter of 12.5 mm at 18 to 21 days. When the corpus luteum regresses, it no longer is producing progestins. It loses its yellowish color, eventually appearing as a small white scar on the surface of the ovary, which is called a *corpus albican*. If the animal is pregnant, the corpus luteum will not regress until late pregnancy for most species, but species differences do exist (see Section 8-5).

**2-2
Oviducts**

The *oviducts* (also called *fallopian tubes*) are a pair of convoluted tubes extending from near the ovaries to and becoming continuous with the tips of the uterine horns. Their functions include transport of ova and spermatozoa which must be conveyed in opposite directions. In addition they are the site of fertilization and the early cell divisions of the embryo. Histologically, they contain three distinct cell layers. The outer layer, basically connective tissue, is the *tunica serosa*. The middle layer, composed of both circular and longitudinal smooth muscle fibers, is the *tunica muscularis*. The innermost layer, which contains both ciliated and secretory epithelial cells is the *tunica mucosa*. The same basic histological arrangement is found in the remainder of the female duct system with some differences in the inner two layers, which will be noted when discussing specific organs.

An oviduct, which is from 20 to 30 cm long for most farm species, is divided into three segments (Figure 2-5). The funnel-shaped opening near the ovaries is the *infundibulum*. In some species (cat, rabbit, mink, and

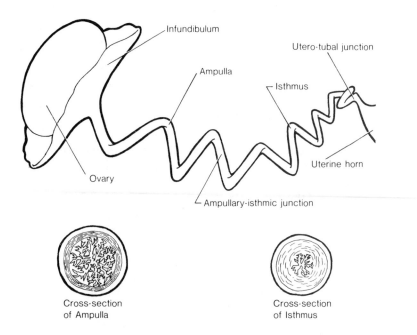

**FIGURE 2-5.**   Anatomy of the oviduct: *Top,* longitudinal view illustrating the macroscopic features of the oviduct; *Bottom,* cross-section of the ampulla and isthmus comparing the thickenss of the musculature of the wall and the complexity of the mucosal folds.

others) the infundibulum forms a bursa around the ovary. In the cow, ewe, sow, and mare the infundibulum is separate from the ovary. There are numerous folds in the mucosa and most mucosal cells in the infundibulum are ciliated. The *ampulla,* the middle segment, is from 3 to 5 mm in diameter and accounts for about half of the total length of the oviduct. The mucosal lining of the ampulla has from 20 to 40 longitudinal folds, which greatly increase the surface area of the lumen. The majority of the cells in the mucosa of the ampulla are ciliated, but some secretory cells are present. The ampulla joins the *isthmus,* the third segment, at the *ampullary-isthmic junction.* This junction is difficult to locate anatomically and has been described as a physiological stricture which delays the ovum several hours during transport. Fertilization occurs at this junction. The isthmus is smaller than the ampulla, being 0.5 to 1 mm in diameter. It is further distinguished by having a thicker smooth muscle layer than the ampulla and from 4 to 8 mucosal folds. A higher ratio of secretory to ciliated cells is characteristic of the isthmus. The isthmus joins the tip of the uterine horn at the *uterotubal junction.* In general oviductal activity is stimulated by estrogens and inhibited by progestins.

**2-3**
**Uterus**

The *uterus* extends from the uterotubal junctions to the cervix. For the cow, sow, and mare the overall length may range from 35 to 50 cm. For the sow and cow the uterine horns account for 80 to 90% of the total length, while in the mare the uterine horns account for about 50% of the total length. The uterus of the ewe is less than half the size of the other mentioned species. The major function of the uterus is to retain and nourish the embryo, or fetus. Before the embryo becomes attached to the uterus, the nourishment comes from *yolk* within the embryo or from *uterine milk* which is secreted by glands in the mucosal layer of the uterus. After attachment to the uterus, nutrients and waste products are conveyed between maternal and embryonic, or fetal, blood by way of the placenta (see Chapter 8).

Four basic types of uteri are found in animals (Figure 2-6). Only two of these types are found in farm animals. The *bicornuate uterus* is found in the sow, cow, and ewe. It is characterized by a small uterine body just anterior to the cervical canal and two long uterine horns. Fusion of the uterine horns of the cow and ewe near the uterine body gives the impression of a larger uterine body than actually exists and has sometimes resulted in their uteri being classified bipartite. The sow has longer uterine horns than the cow and they may be slightly convoluted. The mare has a *bipartite uterus*. There is a prominent uterine body anterior to the cervical canal and two uterine horns that are not as long and distinct as in the bicornuate type. The *duplex uterus*, which consists of two uterine horns each with a separate cervical canal which opens into the vagina, is found in the rat, rabbit, guinea pig, and other small

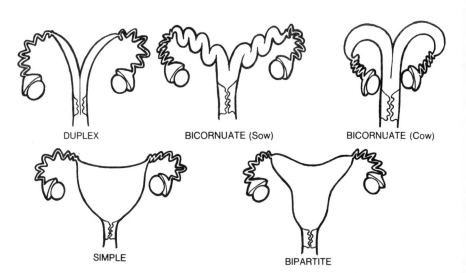

DUPLEX                    BICORNUATE (Sow)                    BICORNUATE (Cow)

SIMPLE                    BIPARTITE

FIGURE 2-6. Basic types of uteri found in mammals.

animals. The *simple uterus,* a pear-shaped body with no uterine horns, is characteristic of humans and other primates.

As in the oviduct, the tunica serosa is the outer layer of the uterus. The *myometrium,* the middle layer, is composed of two thin longitudinal layers of smooth muscle, with a thicker circular layer sandwiched between. Estrogens increase the tone of the myometrium, giving the uterus an "erect" feel. Progestins decrease the tone of the myometrium, causing the uterus to feel more flaccid. The *endometrium,* the mucosal lining of the uterus, is more complex than the rest of the duct system and has simple glands. Estrogens increase the vascularity and cause a thickening of the endometrium. In addition, estrogens stimulate growth of the endometrial glands. Progestins cause the endometrial glands to coil, branch, and secrete uterine milk. The synergistic actions of estrogens and progestins on the endometrium are for preparation of the uterus for pregnancy.

The endometrium provides a mechanism for attachment of the ex-traembryonic membranes. This union forms the placenta, and the process is called *placentation.* With formation of the placenta, nutrients from maternal blood can be transferred to embryonic, or fetal, blood and waste products from embryonic blood can be eliminated through the maternal systems. The nature of the placental attachment differs among species (Figures 2-7 and 2-8). Both cows and ewes have *cotyledonary* placental attachments. *Chorionic villi* from the extraembryonic membranes penetrate into *carun-cles* which are button-like projections on the endometrium. This union, chorionic villi and caruncle, forms the *placentome* (also called the cotyledon). There are 70 to 100 such cotyledonary attachments in a pregnant cow in late pregnancy. The sow and mare have a diffuse (surface) placental attachment. The extraembryonic membranes lie in folds on the endometrium, with

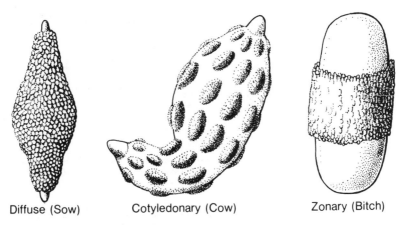

Diffuse (Sow)          Cotyledonary (Cow)          Zonary (Bitch)

FIGURE 2-7.   Distribution of chorionic villi which serve as a basis of placental shape in several species. (Redrawn from Arey. 1974. *Developmental Anatomy.* (7th ed.) W. B. Saunders Co.)

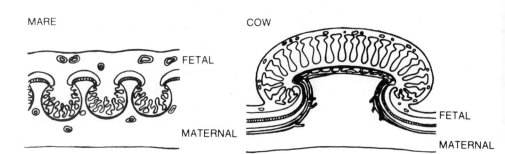

FIGURE 2-8. Diffuse attachment found in the mare and cotyledonary attachment found in the cow.

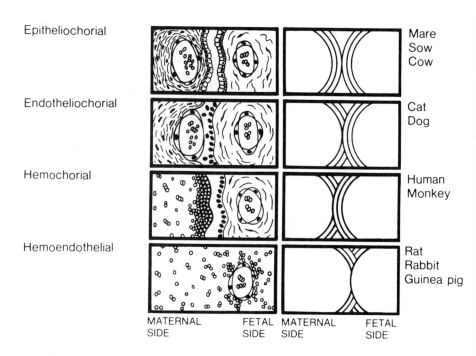

FIGURE 2-9. Placental types showing the cellular barriers between maternal and fetal blood for several species. (Adapted from Arey. 1974. *Developmental Anatomy.* (7th ed.) W. B. Saunders Co.)

chorionic villi extending into the endometrium in a more fragile attachment than is found in the cow or ewe. The placental attachment of the mare, sow, and cow is classified as *epitheliochorial*. This means that no erosion occurs either in the tissues of the extraembryonic membranes or the endometrium during formation of the placenta. Nutrients and gases from the maternal blood must pass through both maternal and extraembryonic tissue layers to reach embryonic blood and vice versa. The ewe has a *syndesmochorial* placental attachment. There is erosion in the epithelial layer of the endometrium. Formation of the placenta in the human results in rather extensive erosion of the endometrium. Classified as *hemochorial*, nutrients from maternal blood must pass through only extraembryonic tissue layers to reach fetal blood. The placental attachment of the rabbit is classified as *hemoendothelial*, with erosion of both endometrial and extraembryonic tissues. Erosion is not extensive enough to result in direct mixing of maternal and fetal blood in any mammalian species (see Figure 2-9).

**2-4**
**Cervix**

While technically a part of the uterus, the *cervix* will be discussed as a distinct organ. It is thick walled and inelastic, the anterior end being continuous with the body of the uterus while the posterior end protrudes into the vagina. For most farm species the length will range from 5 to 10 cm with an outside diameter of 2 to 5 cm. It contains the canal which is the opening into the uterus. The primary function of the cervix is to prevent microbial contamination of the uterus; however, it also may serve as a sperm reservoir after mating. Semen is deposited into the cervix during natural mating in sows and mares.

The cervical canal in the cow and ewe have transverse interlocking ridges known as *annular rings* (Figure 2-10) which help seal the uterus from contaminants. The cervical canal in the sow is funnel-shaped, with ridges in the canal having a corkscrew configuration which conforms to that of the glans penis in the boar (see Chapter 3). The cervical canal of the mare is more open than in other farm species, but mucosal folds in the canal which project into the vagina help prevent contamination.

Histologically, the outer layer of the cervix is the tunica serosa. The middle layer is connective tissue interspersed with smooth muscle fibers, which gives the cervix its firm and inelastic properties. The inner layer, the mucosa, is composed mainly of secretory epithelial cells, but some ciliated epithelial cells are present. High levels of estrogens cause the cervical canal to dilate during estrus (standing heat). Synergism between high levels of estrogens and relaxin cause greater dilation just before parturition. This dilation of the canal would appear to make the uterus more vulnerable to invading organisms. However, estrogens cause the epithelial cells of the cervix to secrete mucus that has antibacterial properties, thus protecting the

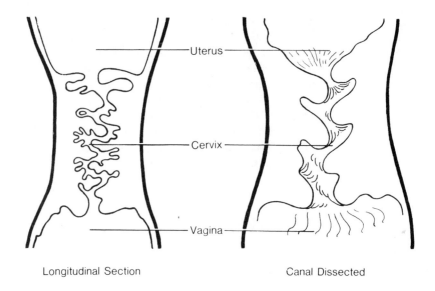

Longitudinal Section                    Canal Dissected

FIGURE 2-10.   The cervix of the cow demonstrating the relationship of the annular rings. (Redrawn from Hafez. 1974. *Reproduction in Farm Animals.* (3rd ed.)
Lea and Febiger.)

uterus. During pregnancy the mucus thickens into a gel-like plug, which
seals and protects the uterus during the pregnancy. Removal of the *mucous
plug* will increase the chance of abortion.

**2-5
Vagina**

The *vagina* is tubular in shape, thin-walled and quite elastic. It is from
25 to 30 cm in length in the cow and mare, and 10 to 15 cm in length in the
sow and ewe. In the cow and ewe, semen is deposited into the anterior end
of the vagina, near the opening to the cervix, during natural mating. The
vagina is the female organ of copulation.

The outer layer, the tunica serosa, is followed by a smooth muscle layer
containing both circular and longitudinal fibers. In most species the mucosal
layer is composed of stratified squamous epithelial cells (the cow a possible
exception). These epithelial cells cornify (become cells without nuclei) under
the influence of estrogens. Vaginal smears can be used as an aid in detecting
estrus, but are most useful in laboratory animals. This layer of cornified cells
at the time of estrus may serve as a lubricating or protective mechanism
which prevents abrasions during copulation. Under the influence of progestins, the epithelial lining regenerates.

**2-6**
**Vulva**

The *vulva,* or external genitalia, consists of the vestibule with related parts and the labia. The *vestibule* is that portion of the female duct system that is common to both the reproductive and urinary systems. It is from 10 to 12 cm in length in cow and mare, half that length in the sow and one-quarter that length in the ewe. The vestibule joins the vagina at the *external urethral orifice.* A *hymen* (ridge) at that point is well defined in the ewe and mare, but less prominent in the cow and sow. A *suburethral diverticulum* (blind pocket) is located just posterior to the external urethral orifice. The labia consist of the *labia minora,* inner folds or lips of the vulva, and *labia majora,* outer folds or lips of the vulva. The labia minora is homologous to the prepuce (sheath) in the male and is not prominent in farm animals. The labia majora, homologous to the scrotum in the male, is that portion of the female system which is visible externally. In the cow the labia majora is covered with fine hair up to the mucosa. The *clitoris,* homologous to the glans penis in the male, is located ventrally and about one cm inside the labia. It contains erectile tissue and is well supplied with sensory nerves. It is erect during estrus. While not prominent enough to be used in estrus detection in most species, the clitoris of the mare is an exception. In the mare during estrus, frequent contractions of the labia (winking) expose the erected clitoris. *Vestibular glands,* located in the posterior part of the vestibule, are active during estrus and secrete a lubricating mucus. The activity of these glands accounts for the moist appearance of the vulva of the cow during estrus.

**2-7**
**Support**
**structures,**
**nerves, and**
**blood supply**

Even though the female reproductive tract may be partially resting on the floor of the pelvis, the *broad ligament* is considered the principal supporting structure. This ligament suspends the ovaries, oviducts, and uterus from either side of the dorsal wall of the pelvis. Blood vessels and nerves pass through the broad ligament to the female reproductive system. The female reproductive system is supplied primarily with autonomic nerves. However, sensory nerves are found in the region of the vulva, especially the clitoral region.

The *ovarian arteries,* also called utero-ovarian arteries, branch and supply blood to the ovaries, oviducts, and a portion of the uterine horns. These arteries are larger on the side of the ovary containing an active corpus luteum in cows and species where one active corpus luteum is the rule. The *middle uterine artery* supplies blood to the rest of the uterine horns and the body of the uterus. It enlarges during middle and late pregnancy and can be palpated as an aid in pregnancy diagnosis in cows and mares (see Chapter 20). The *hypogastric artery* branches to supply the cervix, vagina, and vulva.

Interest in the circulatory patterns of the reproductive tract has increased in recent years since the discovery that the uterus, by release of prostaglandin F$_2\alpha$ (PGF$_2\alpha$), controls the life of the corpus luteum. *Prostag-*

*landin F₂α* is luteolytic (causes the regression of the corpus luteum) but is readily oxidized and essentially destroyed during one passage through the lungs. It did not seem likely that PGF2α, if released into the systemic circulation (uterus→veins→heart and lungs→arteries→ovaries), could be responsible for luteolysis (regression of the corpus luteum). Now there is evidence of a counter current circulation pattern, whereby PGF2α diffuses from the *utero-ovarian vein* into the ovarian artery, thus reaching the ovary by a local rather than a systemic route. A common utero-ovarian vein drains blood from the ovary, oviduct, and a large portion of the uterine horn. In the sow, ewe, and cow the ovarian artery is in close apposition to the utero-ovarian vein (Figure 2-10). In the ewe and cow it is very tortuous, increasing the area of contact with the utero-ovarian vein. The arterial walls are thinnest where they are in contact with this vein. Therefore, it seems likely that sufficient PGF2α can diffuse from the utero-ovarian vein into the ovarian artery to cause luteolysis in the sow, ewe, and cow. During synchronization of estrus, the effective dose of PGF2α is much smaller when infused into the uterus as compared to a systemic injection (5 versus 30 mg).

The ovarian artery of the mare is straight and caudal to the utero-ovarian vein (Figure 2-11). They are in contact in a limited area where the ovarian artery crosses the utero-ovarian vein. Thus a local route for PGF2α from the uterus to the ovary seems less likely in mares than in other species.

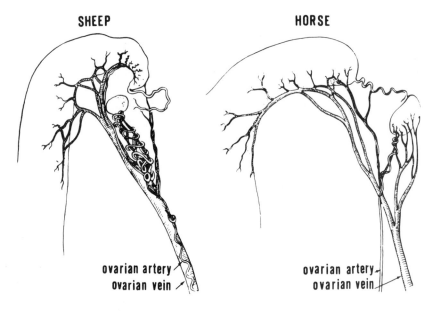

FIGURE 2-11. Comparative diagram of the arteries and veins of a uterine horn and adjacent ovary in the ewe and the mare. (Del Campo and Ginther. 1973. *Amer. J. Vet. Res.,* 34:305.)

**Suggested reading**

Arey, L. B. 1974. *Developmental Anatomy* (7th ed.) W. B. Saunders Co.

Asdell, S. A. 1946. *Patterns of Mammalian Reproduction.* Comstock Pub. Assoc.

Cole, H. H. 1930. A study of the genital tract of the cow with special reference to the cyclic changes. *Amer. J. Anat.*, 46:261.

Del Campo, C. H. and O. J. Ginther. 1973. Vascular anatomy of the uterus and ovaries: Horse, sheep and swine. *Amer. J. Vet. Res.*, 34:305.

Ginther, O. J. and C. H. Del Campo. 1974. Vascular anatomy of the uterus and ovaries and the unilateral luteolytic effect of the uterus: Cattle. *Amer. J. Vet. Res.*, 35:193.

Hafez, E. S. E. 1974. Functional anatomy of the female. *Reproduction in Farm Animals.* (3rd ed.) Lea and Febiger.

Parkes, A. S., ed. 1956. *Marshall's Physiology of Reproduction.* Vol. 1: Part 1. Longmans, Green and Co.

Sisson, S. and J. D. Grossman. 1947. *The Anatomy of Domestic Animals.* W. B. Saunders Co.

# The male
# reproductive system

The male reproductive system (Figure 3-1) consists of the scrotum, spermatic cords, testes, accessory glands, penis, prepuce, and the male duct system. The duct system includes vasa efferentia located within the testis along with the epididymis, vas deferens, and urethra external to the testis. The embryonic origin of the testes is the primary sex cords of the genital ridge, whereas the male duct system originates from the Wolffian ducts. A summary of the reproductive organs of the male and their major functions are listed in Table 3-1.

**3-1**
**Testes**

The *testes* are the primary organs of reproduction in males, just as ovaries are primary organs of reproduction in females. Testes are considered primary because they produce male gametes (spermatozoa) and male sex hormones (androgens). Testes differ from ovaries in that all potential gametes are not present at birth. Germ cells, located in the seminiferous tubules, undergo continual cell divisions forming new spermatozoa throughout the normal reproductive life of the male.

Testes also differ from ovaries in that they do not remain in the body cavity. They descend from their site of origin, near the kidneys, down through the inguinal canals into the scrotum. Descent of the testes occurs because of an apparent shortening of the *gubernaculum*, a ligament extending from the inguinal region and attaching to the tail of the epididymis. This apparent shortening occurs because the gubernaculum does not grow as rapidly as the body wall. The testes are drawn closer to the inguinal canals and intra-abdominal pressure aids passage of the testes through the inguinal canals into the scrotum. Both gonadotrophic hormone and androgens regulate descent of the testes. This descent is completed in the fetus by mid-pregnancy in cattle and just before birth in horses. In some cases one or both testes fail to descend due to a defect in development. If neither descends, the animal is termed a *bilateral cryptorchid*. Bilateral cryptorchids are sterile (see section 3-2). If only one testis descends it is a *unilateral cryptorchid*. The unilateral cryptorchid is usually fertile due to the descended testis. The cryptorchid condition can be corrected by surgery, but this is not recommended for farm animals (Chapter 23). The condition can be inherited, therefore, surgical correction would result in the perpetuation of an undesirable trait.

TABLE 3-1.   Reproductive organs of the male with their major functions

| Organ | Function(s) |
|---|---|
| Testis | Production of spermatozoa |
| | Production of androgens |
| Scrotum | Support of the testes |
| | Temperature control of the testes |
| | Protection of the testes |
| Spermatic cord | Support of the testes |
| | Temperature control of the testes |
| Epididymis | Concentration of spermatozoa |
| | Storage of spermatozoa |
| | Maturation of spermatozoa |
| | Transport of spermatozoa |
| Vas deferens | Transport of spermatozoa |
| Urethra | Transport of semen |
| Vesicular glands | Contributes fluid, energy substrates, and buffers to semen |
| Prostate gland | Contributes fluid and inorganic ions to semen |
| Bulbourethral glands | Flushes urine residue from urethra |
| Penis | Male organ of copulation |
| Prepuce | Encloses free end of penis |

**3-1.1 Functional morphology**

The testis of the bull is 10 to 13 cm long, 5 to 6.5 cm wide and weighs 300 to 400 gm. The testis is of similar size in boars, but is smaller in rams and stallions.

In all species testes are covered with the *tunica vaginalis*, a serous tissue, which is an extension of the peritoneum. This serous coat is obtained as the testes descend into the scrotum and is attached along the line of the

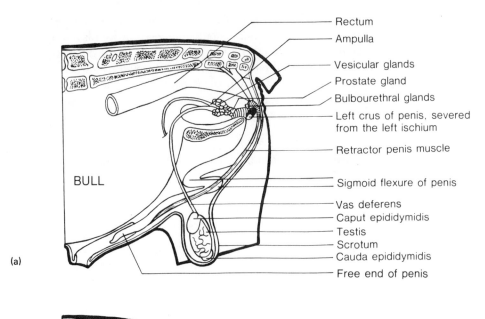

BULL

Rectum
Ampulla
Vesicular glands
Prostate gland
Bulbourethral glands
Left crus of penis, severed from the left ischium
Retractor penis muscle
Sigmoid flexure of penis
Vas deferens
Caput epididymidis
Testis
Scrotum
Cauda epididymidis
Free end of penis

(a)

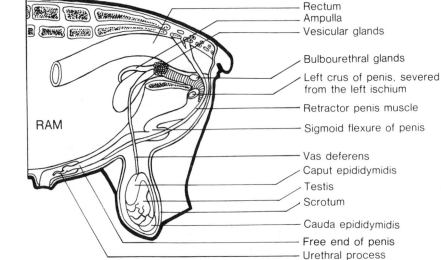

RAM

Rectum
Ampulla
Vesicular glands
Bulbourethral glands
Left crus of penis, severed from the left ischium
Retractor penis muscle
Sigmoid flexure of penis
Vas deferens
Caput epididymidis
Testis
Scrotum
Cauda epididymidis
Free end of penis
Urethral process

(b)

FIGURE 3-1. Diagram of the reproductive system of the bull (a), ram (b), boar (c) and stallion (d). (Redrawn from Sorenson, 1979. *Animal Reproduction: Principles and Practices.* McGraw-Hill).

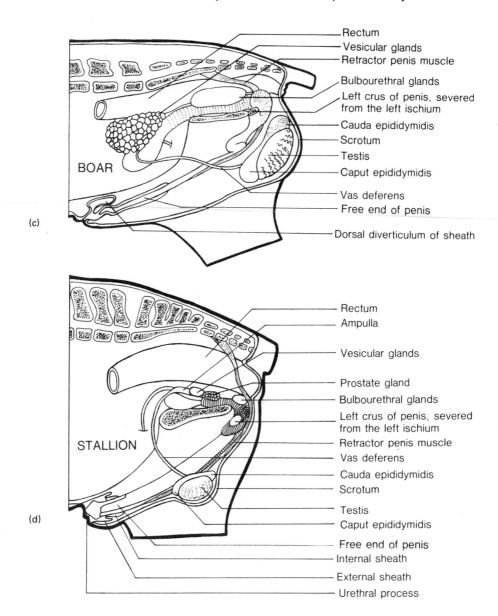

(c)

BOAR

Rectum
Vesicular glands
Retractor penis muscle
Bulbourethral glands
Left crus of penis, severed from the left ischium
Cauda epididymidis
Scrotum
Testis
Caput epididymidis
Vas deferens
Free end of penis
Dorsal diverticulum of sheath

(d)

STALLION

Rectum
Ampulla
Vesicular glands
Prostate gland
Bulbourethral glands
Left crus of penis, severed from the left ischium
Retractor penis muscle
Vas deferens
Cauda epididymidis
Scrotum
Testis
Caput epididymidis
Free end of penis
Internal sheath
External sheath
Urethral process

epididymis. The outer layer of the testes, the *tunica albuginea testis*, is a thin white membrane of elastic connective tissue. Numerous blood vessels are visible just under its surface. Beneath the tunica albuginea testis is the *parenchyma*, the functional layer of the testes. The parenchyma has a yellowish color and is divided into segments by incomplete septa of connective tissue (Figure 3-2). Located within these segments of parenchymal tissue are the *seminiferous tubules*. Seminiferous tubules are formed from primary sex cords. They contain germ cells (spermatogonia) and nurse cells (Sertoli cells). Sertoli cells are larger and less numerous than spermatogonia. With stimulation by FSH, Sertoli cells produce both androgen binding protein and inhibin (see Chapter 4). Seminiferous tubules are the site of spermatozoa production. They are small, convoluted tubules approximately 200 $\mu$ in diameter. It has been estimated that the seminiferous tubules from a pair of bull testes, stretched out and laid end to end, approach 5 Km in length. They make up 80% of the weight of the testes. Seminiferous tubules join a network of tubules, the *rete testis*, which connects to 12 to 15 small ducts, the *vasa efferentia*, which converges into the head of the epididymis. Production of spermatozoa will be discussed in Chapter 6.

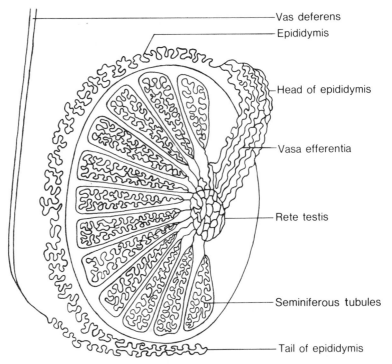

FIGURE 3-2.  Sagittal section of testis illustrating segments of parenchymal tissue which contain the seminiferous tubules, rete testis, vasa efferentia, epididymis, and scrotal portion of the vas deferens.

Leydig (interstitial) cells are found in the parenchyma of the testes between the seminiferous tubules (Figure 3-3). LH stimulates Leydig cells to produce testosterone and small quantities of other androgens.

Testosterone is needed for development of secondary sex characteristics and for normal mating behavior. In addition, it is necessary for the function of the accessory glands, production of spermatozoa and maintenance of the male duct system. Through its effects on the male, testosterone aids in maintenance of optimum conditions for spermatogenesis, transport of spermatozoa and deposition of spermatozoa into the female tract. Normal body temperature will not affect the function of the Leydig cells. For example, bilateral cryptorchids develop secondary sex characteristics, have normal sexual vigor and can do all things associated with reproduction except production of spermatozoa.

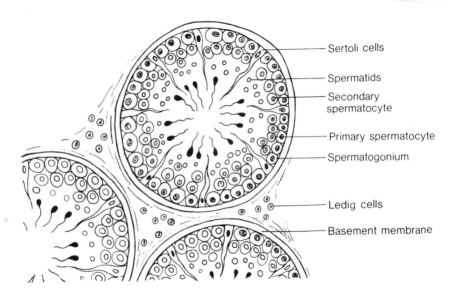

FIGURE 3-3. Cross section of parenchymal tissue showing relationship between the seminiferous tubules and interstitial tissue containing Leydig cells.

**3-2
Scrotum and
spermatic cord**

The *scrotum* is a two-lobed sac which encloses the testes. It is located in the inguinal region between the rear legs of most species. The scrotum has the same embryonic origin as the labia majora in the female. It is composed of an outer layer of thick skin with numerous large sweat and sebaceous glands. This outer layer is lined with a layer of smooth muscle fibers, the *tunica dartos*, which is interspersed with connective tissue. The tunica dartos divides the scrotum into two pouches, and is attached to the tunica vaginalis at the bottom of each pouch.

The *spermatic cord* connects the testis to its life support mechanisms, the convoluted testicular arteries and surrounding venus plexus, and nerve trunks. In addition, the spermatic cord is composed of smooth muscle fibers, connective tissue, and a portion of the vas deferens. Both the spermatic cords and scrotum contribute to the support of the testes. Also, they have a joint function in regulating the temperature of the testes.

## 3-2.1 Temperature control

Several examples can be given to illustrate the importance of temperature control of the testes. If a ram's scrotum is insulated, or the testes are tied against the abdomen, sterility results. The higher temperature causes degeneration of the cells lining the wall of the seminiferous tubules. Fertility will be restored if the testes and scrotum are returned to their natural state before total degeneration occurs. However, a few weeks will be required before fertile semen is again produced. (Sometimes men with high fevers are sterile for short periods after recovery.) The bilateral cryptorchid is sterile, again illustrating that production of spermatozoa stops when the temperature inside the testes is as high as normal body temperature.

Low fertility semen produced by several species of farm animals during the summer has been attributed to the inability of the body's cooling mechanisms to keep the testes cool enough. In cattle, when ambient temperatures range from 5 to 21°C, temperatures inside of the testes will be 4 to 7°C below body temperature. As ambient temperature increases to approximately 38°C, the differential between body temperature and testes temperature may be reduced by half (2 to 3°C below body temperature). There is no evidence that low ambient temperature will lower fertility.

The role of the scrotum and spermatic cord in temperature control of the testes involves drawing the testes closer to the body as ambient temperature falls and letting the testes swing further away from the body as ambient temperature rises. Two smooth muscles are involved. The *tunica dartos*, the smooth muscle lining the scrotum, and the *cremaster*, a smooth muscle around the spermatic cord, are sensitive to temperature. During cold weather, contraction of these muscles causes the scrotum to pucker and the spermatic cords to shorten, drawing the testes closer to the body. During hot weather, these muscles relax permitting the scrotum to stretch and the spermatic cord to lengthen. Thus, the testes swing down away from the body. These muscles do not respond to changes in temperature until near the age of puberty. They must be sensitized by testosterone to respond to changing ambient temperature.

Actual cooling of the testes occurs by two mechanisms. The skin of the scrotum has both sweat and sebaceous glands which are more active during hot weather. Evaporation of the secretions of these glands cools the scrotum and thus the testes. The external scrotum has been observed to be 2 to 5°C cooler than the temperature inside the testes. As the scrotum stretches during hot weather, more surface area is provided for cooling by evaporation. In addition to cooling by evaporation, significant cooling occurs through heat exchange in the circulatory system (Figure 3-4). As arteries transporting blood at internal body temperature transcend the spermatic cord, their convoluted folds pass through a network of veins, the *pampiniform venous plexus,* transporting cooler blood back towards the heart. Some cooling of arterial blood then occurs before it reaches the testes. The lengthening of the cord during hot weather provides more surface area for this heat exchange.

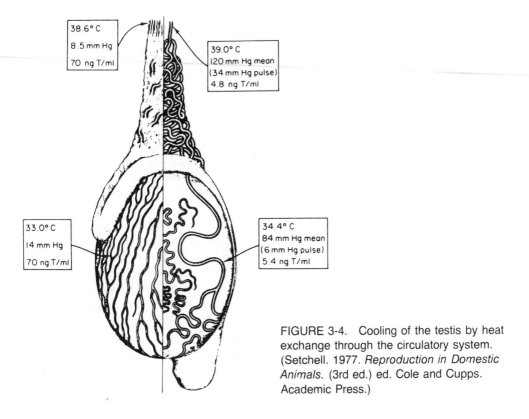

FIGURE 3-4. Cooling of the testis by heat exchange through the circulatory system. (Setchell. 1977. *Reproduction in Domestic Animals.* (3rd ed.) ed. Cole and Cupps. Academic Press.)

**3-3**
**Epididymis**

The *epididymis,* the first external duct leading from the testis, is fused longitudinally to the surface of the testis and is encased in the tunica vaginalis with the testis. The single convoluted duct is covered with an extension of the tunica albuginea testis (Figure 3-5). The *caput* (head) of the epididymis is a flattened area at the apex of the testis where 12 to 15 small ducts, the vasa efferentia, merge into a single duct. The *corpus* (body) extending along the longitudinal axis of the testis, is a single duct which becomes continuous with the *cauda* (tail). The total length of this convoluted duct is about 34 meters in the bull and is longer in the ram, boar, and stallion. The lumen of the cauda is wider than the lumen of the corpus. The structure of the epididymis and other external ducts (vas deferens and urethra) is similar to that of the tubular portion of the female tract. The tunica serosa (outer layer) is followed by a smooth muscle layer (middle) and an epithelial layer (innermost).

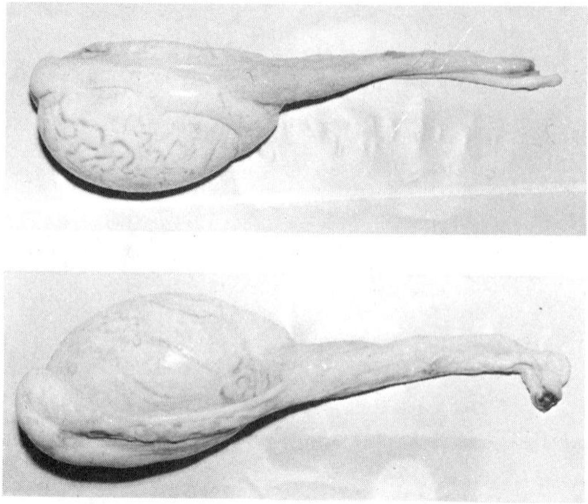

FIGURE 3-5.   View of the epididymis as it is fused to the surface of the testis. Note the tunica albuginea testis which covers the testis.

**3-3.1
Transport**

As a duct leading from the testes, the epididymis serves to transport spermatozoa. In sexually active males the time involved in transport is 9 to 14 days in boars, 13 to 15 days in rams, and 9 to 11 days in bulls. Frequent ejaculation has been reported to speed transport by 10 to 20%.

Several factors contribute to movement of spermatozoa through the epididymis. One factor is pressure from the production of new spermatozoa. As spermatozoa are produced in the seminiferous tubules, they are forced out through the rete testis and vasa efferentia into the epididymis. In a sexually inactive male, they are eventually forced through the epididymis. Such movement of spermatozoa is aided by external pressure created by the massaging effect on the testes and epididymis that occurs during normal exercise. The lining of the epididymis contains some ciliated epithelial cells but the role of these cilia in facilitating movement of spermatozoa is not clear. As mentioned previously, movement of spermatozoa is aided by ejaculation. During ejaculation, peristaltic contractions involving the smooth muscle layer of the epididymis and a slight negative pressure (sucking action) created by peristaltic contractions of the vas deferens and urethra actively move spermatozoa from the epididymis into the vas deferens and urethra.

**3-3.2
Concentration**

A second function of the epididymis is concentration of spermatozoa. Spermatozoa entering the epididymis from the testis of the bull, ram, and boar are relatively dilute (approximately 100,000,000 spermatozoa/cc). In the epididymis, they concentrate to about $4 \times 10^9$ (4 billion) spermatozoa per cc. Concentration occurs as the fluids, which suspend spermatozoa in the testes, are absorbed by the epithelial cells of the epididymis. Absorption of these fluids occurs principally in the caput and proximal end of the corpus.

**3-3.3**
**Storage**

A third function of the epididymis is storage of spermatozoa. Most are stored in the cauda of the epididymis where concentrated spermatozoa are packed into the wide lumen. The epididymis of a mature bull may contain 50 to 74 billion sperm. Capacities of other species are not reported. Conditions are optimum in the cauda for preserving the viability of spermatozoa for an extended period. The low pH, high viscosity, high carbon dioxide concentration, high potassium-to-sodium ratio, the influence of testosterone, and probably other factors combine to contribute to a lower metabolic rate and extended life. These conditions have not been duplicated outside the epididymis. If the epididymis is ligated to prevent entry of new spermatoza and removal of old, spermatozoa have remained alive and fertile for about 60 days. On the other hand, after a long period of sexual rest, the first few ejaculates may contain a high percentage of nonfertile spermatozoa.

**3-3.4**
**Maturation**

A fourth function of the epididymis is that of maturation of spermatozoa. When recently formed spermatozoa enter the caput from the vasa efferentia they have the ability for neither motility nor fertility. As they pass through the epididymis they gain the ability to be both motile and fertile. If the cauda is ligated at each end, those spermatozoa closest to the corpus have increased in fertility for up to 25 days. During the same period, those closest to the vas deferens exhibited reduced fertilizing ability. Therefore, it appears that spermatozoa gain ability to be fertile in the cauda and then start to age and deteriorate if they are not removed.

While in the epididymis, spermatozoa lose the cytoplasmic droplet which forms on the neck of each spermatozoa during spermatogenesis (see Chapter 6). The physiological significance of the cytoplasmic droplet is not known, but it has been used as an indicator of sufficient maturation of spermatozoa in the epididymis. If a high percentage of spermatozoa in freshly ejaculated semen has cytoplasmic droplets, they are considered immature and have low fertilizing capacity.

**3-4**
**Vas deferens**
**and urethra**

The *vas deferens* is a duct with one leading from the distal end of the cauda of each epididymis. Initially supported by folds of the peritoneum, it then passes along the spermatic cord, through the inguinal canal to the pelvic region where it merges with the urethra at its origin near the opening to the bladder. The enlarged end of the vas deferens near the urethra is the *ampulla*. The vas deferens has a thick layer of smooth muscles in its walls and appears to have the single function of transport of spermatozoa. Some have suggested that ampullae serve as a short term storage depot for semen. However, spermatozoa age quickly in the ampullae. It seems more likely that spermatozoa may pool in the ampullae during ejaculation before being expelled into the urethra.

The urethra is a single duct which extends from the junction of the ampullae to the end of the penis. It serves as an excretory duct for both urine and semen. During ejaculation in the bull and ram there is a complete mixing of spermatozoa concentrate from the vas deferens and epididymis with fluids from the accessory glands in the pelvic part of the urethra to form semen. In stallions and boars, mixing is not as complete, with the ejaculate containing sperm-free and sperm-rich segments (see Chapter 11).

**3-5
Accessory
glands**

The *accessory glands* (Figure 3-6) are located along the pelvic portion of the urethra, with ducts which empty their secretions into the urethra. They include the *vesicular glands, prostate gland,* and the *bulbourethral glands.* They contribute greatly to the fluid volume of semen. In addition, their secretions are a solution of buffers, nutrients, and other substances needed to assure optimum motility and fertility of semen.

**3-5.1
Vesicular
glands**

The vesicular glands (sometimes called seminal vesicles) are a pair of lobular glands that are easily identified because of their knobby appearance. They have been described as having the appearance of a "cluster of grapes." They are of similar length in the bull, boar, and stallion (13 to 15 cm), but the width and thickness of the vesicular glands of the bull is approximately half that of the boar and stallion. The vesicular glands of the ram are much smaller, being about 4 cm in length. The excretory ducts of the vesicular glands open near the bifurcation where the ampullae merge with the urethra. In bulls, they contribute well over half of the total fluid volume of semen, and appear to make a substantial contribution in other species. Several organic compounds found in secretions of the vesicular glands are unique in that they are not found in substantial quantities elsewhere in the body. Two of these compounds, fructose and sorbitol, are major sources of energy for bull and ram spermatozoa but are found in lower concentration in boar and stallion semen. Both phosphate and carbonate buffers are found in these secretions and are important in that they protect against shifts in the pH of semen. Such shifts in pH would be detrimental to spermatozoa.

**3-5.2
Prostate gland**

The prostate is a single gland located around and along the urethra just posterior to the excretory ducts of the vesicular glands. A prostate body is visible in excised tracts and can be palpated in bulls and stallions. In rams, all of the prostate is embedded in urethral muscles as is part of this glandular tissue in bulls and boars. It makes a small contribution to the fluid volume of semen in most species studied. However, some report that the contribution

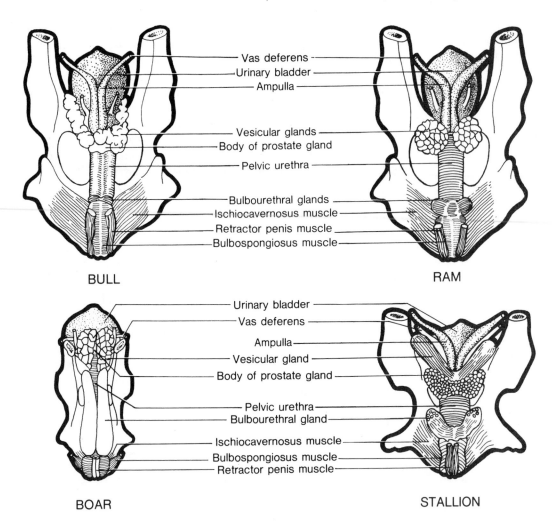

Vas deferens
Urinary bladder
Ampulla

Vesicular glands
Body of prostate gland
Pelvic urethra

Bulbourethral glands
Ischiocavernosus muscle
Retractor penis muscle
Bulbospongiosus muscle

BULL

RAM

Urinary bladder
Vas deferens
Ampulla
Vesicular gland
Body of prostate gland

Pelvic urethra
Bulbourethral gland

Ischiocavernosus muscle
Bulbospongiosus muscle
Retractor penis muscle

BOAR

STALLION

FIGURE 3-6. Accessory glands of the bull, boar, ram, and stallion showing their relationship to the ampulla and urethra. (Redrawn from Ashdown and Hancock. 1974. *Reproduction in Farm Animals*. (3rd ed.) ed. Hafez. Lea and Febiger.)

of the prostate gland is at least as substantial as that of the vesicular glands in boars. The prostate of the boar is larger than that of the bull. The secretions of the prostate are high in inorganic ions with sodium, chlorine, calcium, and magnesium all in solution.

**3-5.3
Bulbourethral
glands**

The bulbourethral (Cowpers) glands are a pair of glands located along the urethra near the point where it exits from the pelvis. They are about the size and shape of walnuts in bulls, but are much larger in boars. In bulls, they are embedded in the bulbospongiosum muscle. They contribute very little to the fluid volume of semen. In bulls, their secretions flush urine residue from the urethra before ejaculation. These secretions are seen as dribblings from the prepuce just before copulation. In boars, their secretions account for that portion of boar semen which coagulates. This is strained from boar semen before it is used for artificial insemination. During natural service, the white lumps formed by coagulation may prevent semen from flowing back through the cervix into the vagina of sows.

**3-6
Penis**

The *penis* is the organ of copulation in males (Figure 3-1). It forms dorsally around the urethra from the point where the urethra leaves the pelvis, with the external urethral orifice at the free end of the penis. Bulls, boars, and rams have a *sigmoid flexure*, an S-shaped bend in the penis which permits it to be retracted completely into the body. These three species and the stallion have *retractor penis muscles*, a pair of smooth muscles which will relax to permit extension of the penis and contract to draw the penis back into the body. These retractor penis muscles arise from the vertebrae in the coccygeal region and are fused to the ventral penis just anterior to the sigmoid flexure. The *glans penis* (Figure 3-7), which is the free end of the penis, is well supplied with sensory nerves and is homologous to the clitoris in the female (Chapter 2). In most species the glans penis is fibroelastic,

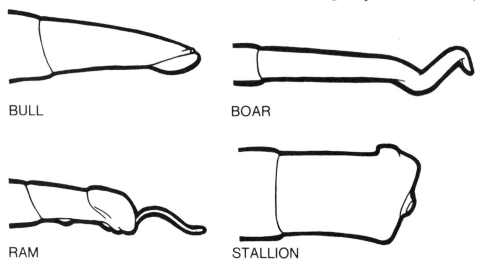

BULL　　　　　　　　　　BOAR

RAM　　　　　　　　　　STALLION

FIGURE 3-7. Comparative diagram showing the shape of the glans penis of the bull, boar, ram, and stallion. (Redrawn from Ashdown and Hancock. 1974. *Reproduction in Farm Animals.* (3rd ed.) ed. Hafez. Lea and Febiger.)

containing small amounts of erectile tissue. The glans penis of stallions contains more erectile tissue than is found in bulls, boars, or rams.

Erectile tissue is cavernous (spongy) tissue located in two regions of the penis (Figure 3-8). The *corpus spongiosum penis* is the cavernous tissue around the urethra. It enlarges into the penile bulb which is covered with *bulbospongiosum* muscle at the base of the penis. The *corpus cavernosum penis* is a larger area of cavernous tissue located dorsally to the corpus spongiosum penis. It arises as two cavernous rods from the *ischiocavernosus muscle,* eventually fusing to form one cavernous area as it proceeds toward the glans penis. These cavernous areas engorge with blood during sexual excitement causing extension of the penis (erection) and facilitating the final ejection of semen during ejaculation (Chapter 11). Both the bulbospongiosum muscle and ischiocavernosus muscle are striated, skeletal muscles, rather than the smooth muscle associated with most of the male and female tracts.

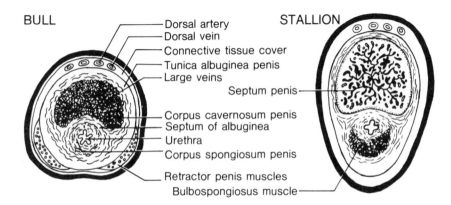

FIGURE 3-8.   Cross section of penis showing corpus cavernosum penis and corpus spongiosum penis. (Redrawn from Sorenson. 1979. *Animal Reproduction: Principles and Practices.* McGraw-Hill.)

**3-7
Prepuce**

The *prepuce* (sheath) is an invagination of skin which completely encloses the free end of the penis. It has the same embryonic origin as the labia minora in the female. It can be divided into a prepenile portion which is the outer fold and the penile portion or inner folds. The orifice of the prepuce is surrounded by long and tough preputial hairs.

**Suggested
reading**

Arey, L.B. 1947. *Developmental Anatomy.* (5th ed.) W. B. Saunders Co.

Ashdown, R. R. and J. L. Hancock. 1974. Functional anatomy of male reproduction. *Reproduction in Farm Animals.* (3rd ed.) ed. E. S. E. Hafez. Lea and Febiger.

Johnson, A. D., W. R. Gomes and N. L. VanDemark. 1970. *The Testes.* Vol. 1. Academic Press.

Parkes, A. S. 1956. *Marshall's Physiology of Reproduction.* Vol. 1. Part 1. Longmans, Green and Co.

Salisbury, G. W., N. L. VanDemark and J. R. Lodge. 1978. The reproductive tract of the bull. *Physiology of Reproduction and Artificial Insemination of Cattle.* (2nd ed.) W. H. Freeman and Co.

Setchell, B. P. 1977. Male reproductive organs and semen. *Reproduction in Domestic Animals.* (3rd ed.) ed. H. H. Cole and P. T. Cupps. Academic Press.

Sisson, S. and J. D. Grossman. 1947. *The Anatomy of Domestic Animals.* W. B. Saunders Co.

# Chapter

# 4

# Natural synchronization processes

Before undertaking a study of reproduction it is important to become familiar with the physiological system most responsible for regulating the natural reproductive processes. The endocrine system (Figure 4-1), through the hormones that it produces, is responsible for this regulation.

*Endocrine glands* are ductless glands and secrete internally. They secrete directly into the blood stream, as opposed to exocrine glands which have ducts and secrete externally, not into the blood stream. Endocrine glands secrete hormones—chemical agents—which are carried by the blood to cells within a target organ or to other target cells where they regulate a specific physiological activity. Thus, hormones exert their influence through cells at a site away from the glands that produce them.

Cells that respond to a specific hormone do so because they have *receptor sites* which bind that hormone. Receptor sites can be defined as recognition units in cells that have a high affinity for a particular hormone. The concentration of receptor sites in a target organ will increase or decrease depending on the endocrine status of the animal. Regulation of hormone receptor sites in organs of reproduction will be discussed in section 4-8. Knowledge of how receptor sites are regulated has added a new realm of understanding to the endocrine control of reproduction.

Chemically, hormones of reproduction can be divided into two classes. One class includes the peptide and protein hormones (Table 4-1). These hormones are formed by the bonding of a series of amino acids, with molecular size being the determinate of whether they are called peptide or protein. Peptide and protein hormones are soluble in water. They are denatured, making them physiologically inactive, by strong acids, strong bases, or by heat. To be physiologically effective they must be administered systemically (i.v., i.m., or s.c.) rather than orally.

The second class of hormones of reproduction are steroids. Steroids are a special class of lipids having a tetracyclic configuration (Figure 4-2). All

37

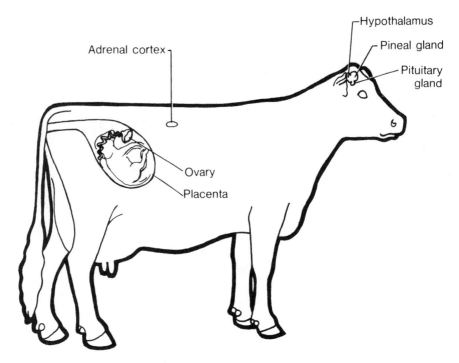

FIGURE 4-1. Approximate location of the endocrine glands of the cow which secrete hormones that regulate reproduction. (Redrawn from Foley *et al.* 1972. *Dairy Cattle: Principles, Practices, Problems, Profits.* Lea and Febiger.)

TABLE 4-1. Molecular size of peptide and protein hormones which regulate reproduction

| Hormone | Molecular wt |
|---|---|
| FSH | 28,000 to 32,000 |
| LH | 26,000 to 30,000 |
| Prolactin | 23,000 to 25,000 |
| ACTH | 4500 |
| Inhibin | > 10,000 |
| Oxytocin | 1007 |
| GnRH | 1200 to 1500 |
| HCG | 37,700 |
| PMSG | 28,000 |
| Relaxin | 6500 |

steroid hormones have cholesterol as a common precurser. They are not soluble in water but are soluble in ether, chloroform, and other solvents which can be used to extract lipids from tissue or blood. Some steroids can be effectively absorbed through the gastrointestinal tract but are usually less effective with oral administration than with systemic administration. Natural estrogens are usually more effective than natural progestins or androgens when given orally. Synthetic progestins have been developed for oral administration.

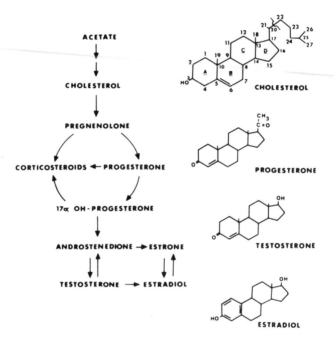

FIGURE 4-2.  Metabolic pathway for the synthesis of gonadal steroid hormones and the chemical structure of the three most important sex steroids. (Niswender *et al.* 1974. *Reproduction in Farm Animals.* (3rd ed.) ed. Hafez. Lea and Febiger.)

Functionally, hormones can be classified as either *primary hormones of reproduction* or *secondary hormones of reproduction*. The primary hormones of reproduction are those which directly regulate a reproductive activity. Most other hormones are classified as secondary hormones of reproduction. They are considered necessary for the maintenance of the proper internal environment for normal reproduction. The discussion in this chapter will be limited to the primary hormones of reproduction (Table 4-2). Functions of hormones which are mentioned briefly in this chapter will be discussed in greater detail in later chapters.

TABLE 4-2. Hormones that regulate reproduction

| Gland | Hormone | Chemical class | Principal functions |
|---|---|---|---|
| **Hypothalamus** | Gonadotrophic releasing hormone | Peptide | FSH and LH release |
| | Prolactin inhibiting hormone | " | Prolactin retention |
| | Prolactin releasing hormone | " | Prolactin release |
| | Corticotrophic releasing hormone | " | ACTH release |
| **Anterior Pituitary** | Follicle stimulating hormone (FSH) | Protein | (1) Follicle growth<br>(2) Estrogen release<br>(3) Spermiogenesis |
| | Luteinizing hormone (LH) | " | (1) Ovulation<br>(2) Corpus luteum formation and function<br>(3) Testosterone release |
| | Prolactin | " | (1) Milk synthesis |
| | Adrenalcorticotrophic hormone (ACTH) | Polypeptide | (1) Release of glucocorticoids |
| **Posterior Pituitary** | Oxytocin | Peptide | (1) Parturition<br>(2) Milk ejection |
| **Ovary** | Estrogens (Estradiol) | Steroid | (1) Mating behavior<br>(2) Secondary sex characteristics<br>(3) Maintenance of female duct system<br>(4) Mammary growth |
| | Progestins (Progesterone) | Steroid | (1) Maintenance of pregnancy<br>(2) Mammary growth |
| | Relaxin | Polypeptide | (1) Expansion of pelvis<br>(2) Dilation of cervix |
| | Inhibin | Protein | (1) Prevents release of FSH |
| **Testis** | Androgens (Testosterone) | Steroid | (1) Male mating behavior<br>(2) Spermatocytogenesis<br>(3) Maintenance of male duct system<br>(4) Function of accessory glands |
| | Inhibin | Protein | (1) Prevents release of FSH |
| **Adrenal Cortex** | Glucocorticoids (Cortisol) | Steroid | (1) Parturition<br>(2) Milk synthesis |

| | | | |
|---|---|---|---|
| **Placenta** | Human chorionic gonadotrophin (HCG) | Protein | (1) LH-like |
| | Pregnant mare serum gonadotrophin (PMSG) | Protein | (1) FSH-like |
| | | | (2) Supplementary corpora lutea in mare |
| | Estrogens Progestins Relaxin | (See ovary) | |
| **Uterus** | Prostaglandin F$_{2\alpha}$ (PGF$_{2\alpha}$) | Lipid | (1) Regression Corpus luteum |
| | | | (2) Parturition |

**4-1**
**Primary reproductive hormones of the pituitary gland**

The pituitary, a gland located in a bony depression (the sella turcica) at the base of the brain, is embryologically and functionally two separate glands in the adult animal. The anterior lobe or *anterior pituitary* (also called adenohypophysis) arises from embryonic gut tissues of the roof of the mouth. The posterior lobe or *posterior pituitary* (also called neurohypophysis) forms from embryonic brain tissue.

The anterior pituitary produces three primary hormones of reproduction. These protein hormones of the anterior pituitary are *follicle stimulating hormone* (FSH), *luteinizing hormone* (LH), and *prolactin*. Collectively, FSH and LH are known as *gonadotrophins* because they stimulate the gonads. In the male, LH has sometimes been called *interstitial cell stimulating hormone* (ICSH).

In the female, FSH promotes follicle growth and estrogen production by the ovaries. Also, FSH stimulates the follicle to produce an inhibin-like protein. *Inhibin* is a protein produced by the testes which acts directly on the anterior pituitary to prevent the release of FSH. Characterization of these proteins is not complete enough to know if they are the same compound. Functionally they appear the same. LH causes maturation of the follicle, *ovulation* (rupture of the follicle and release of ovum) and is luteotrophic. That is, it stimulates formation of the corpus luteum and production of progesterone. Prolactin synergizes with LH by increasing LH receptor sites in the corpus luteum in some species. Also, prolactin has a stimulating effect on development of the mammary gland and the synthesis of milk. *Adrenalcorticotrophic hormone* (ACTH), a small protein hormone of the anterior pituitary, stimulates the release of *glucocorticoids* from the adrenal cortex. Glucocorticoids play a role in parturition and in the synthesis of milk.

In the male, FSH stimulates spermiogenesis in the testes with action on both spermatogonia and Sertoli cells. FSH stimulates the Sertoli cells to produce inhibin and *androgen binding protein* (ABP). ABP is secreted into the lumen of the seminiferous tubules and serves as a carrier for testosterone. LH stimulates the cells of Leydig, located in the interstitial tissue of the testes, to produce testosterone and other androgens. Prolactin appears to synergize with LH by increasing hormone receptor sites for LH in the testes.

*Oxytocin*, a peptide hormone released from the posterior pituitary, stimulates the contraction of smooth muscle in the oviduct and uterus. Because of this activity, it has been postulated that oxytocin aids both sperm and ovum transport in the female tract and stimulates uterine contractions during parturition. Also, oxytocin stimulates the myoepithelial cells of the mammary gland, causing the ejection of milk.

## 4-2 Control of the pituitary gland by the hypothalamus

The *hypothalamus*, which forms the floor and lateral wall of the third ventricle of the brain, is closely linked to the pituitary. The hypophyseal portal blood system connects the hypothalamus with the anterior pituitary, while the posterior pituitary is an extension of the hypothalamus. Nerve fibers from neurosecretory cells in the hypothalamus extend down into the posterior pituitary (Figure 4-3).

Secretion of gonadotrophic hormones from the anterior pituitary is controlled by a peptide-releasing hormone which is produced by neurosecretory cells in the hypothalamus. One peptide, *gonadotrophic releasing hormone* (GnRH), has been isolated and purified from pigs and sheep. GnRH causes the release of both FSH and LH. Until recently, it had been postulated that separate releasing agents (FSH-releasing hormone and LH-releasing hormone) regulated the release of FSH and LH from the anterior pituitary. While some physiological evidence for separate releasing hormones still exists, the preponderance of evidence supports a single releasing hormone for FSH and LH. In a clinical situation, GnRH can be used instead of LH for treatment of cystic ovaries in cows. There is evidence that both a *prolactin-releasing hormone* (PRH) and *prolactin-inhibiting hormone* (PIH) control the release and retention of prolactin in the anterior pituitary. *Corticotrophic-releasing hormone* (CRH) stimulates the release of ACTH. A clearer picture of the functional nature of these releasing hormones should evolve in the near future. It is important that a link has been established between the central nervous system and function of the endocrine system.

Oxytocin, which is released from the posterior pituitary, is produced by neurosecretory cells in the hypothalamus. After synthesis, oxytocin is transported by carrier proteins (neurophysins) as secretory droplets along nerve fibers extending into the posterior pituitary. Stimulation of sensory nerves in the teats or cervix will cause oxytocin to be released from nerve endings in the posterior pituitary.

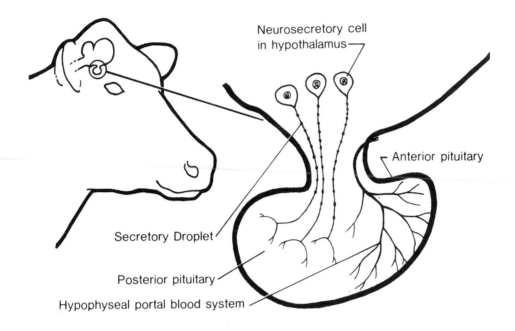

FIGURE 4-3. Relationship between the hypothalamus and the pituitary gland. (The pituitary-hypothalamic area is enlarged to permit a more detailed illustration.)

**4-3**
**Hormones of**
**the gonads**

Major steroid hormones produced by the gonads are shown in Table 4-3.

**4-3.1**
**Female**

Two classes of hormones produced by the ovaries are *estrogens* and *progestins*. Chemically, estrogens and progestins are classified as steroids and have cholesterol as a common precurser.

Estrogens, representing a group of steroids with similar physiological activity, are produced by specific cells in the Graafian follicle. The estrogen of greatest importance, quantitatively and physiologically, is *estradiol*. Others of importance include estriol and estrone. The principal actions of estrogens are their influence on (1) the manifestation of mating behavior during estrus; (2) cyclic changes in the female tract; (3) duct development in the mammary gland; and (4) development of secondary sex characteristics. Estrogens have been called the "female sex hormone." Estrogens are luteolytic in cows and ewes but are luteotrophic in sows.

Progestins are another group of hormones with similar physiological activity, the most important being *progesterone*. They are produced by the

TABLE 4-3. Major steroid hormones produced by the gonads

| Class | Hormone |
|---|---|
| Estrogen | Estradiol-17 $\beta$<br>Estriol<br>Estrone |
| Progestins | Progesterone<br>17-Hydroxyprogesterone<br>20 $\beta$-dihydroprogesterone |
| Androgens | Testosterone<br>Androstenedione<br>Dihydrotestosterone |

corpus luteum. Important functions are (1) inhibition of sexual behavior; (2) maintenance of pregnancy by inhibiting uterine contractions and promoting glandular development in the endometrium; and (3) promotion of alveolar development of the mammary gland. The synergistic actions of estrogens and progestins are notable in preparing the uterus for pregnancy and the mammary gland for lactation.

Both estrogens and progestins help regulate the release of gonado-trophins, acting through both the hypothalamus and anterior pituitary (Figure 4-4). High levels of either progestins or a combination of progestins and estrogens inhibit the release of FSH and LH from the anterior pituitary—a negative feedback control. Near the time of estrus, when progesterone levels are low, high estrogen concentrations stimulate the release of LH and prolactin—a positive feedback control. The influence of the gonadotrophins on estrogen and progestin release has been mentioned previously. Therefore, it can be seen that reciprocal action between the gonadotrophins and the steroid hormones of the ovaries is necessary for maintenance of the hormone balance essential for normal reproduction.

An inhibin-like protein produced by the ovary which selectively suppresses release of FSH, but not LH, from the anterior pituitary helps regulate FSH. This protein may account for some of the reported differences in the release patterns of FSH and LH that appear inconsistent with a single gonadotrophic-releasing hormone.

*Relaxin* is a polypeptide hormone produced by the corpus luteum. Little is known about the mechanisms controlling its production, but higher concentrations are seen during pregnancy. It causes a relaxation of pelvic ligaments and softening of the connective tissue of the uterine muscles to allow the expansion necessary to accommodate the growing fetus. Synergizing with estrogen, it causes further expansion of the pelvis and softening of the connective tissue of the cervix to permit the fetus to be expelled during parturition.

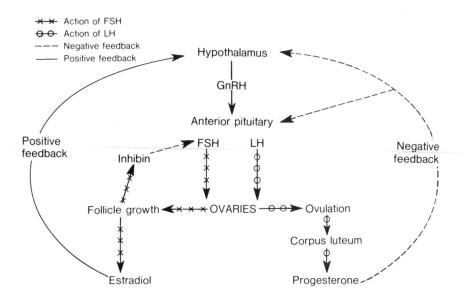

FIGURE 4-4.  Relationship between the hypothalamic releasing hormones, gonadotrophins, and ovarian hormones in regulating reproductive function.

**4-3.2**
**Male**

Upon stimulation by LH, the Leydig cells of the testes produce *androgens*, which are a class of steroid hormones. The principal androgen in mature males is *testosterone*, which has been labeled the male sex hormone. Dihydrotestosterone is found in high enough concentration in peripheral tissue to be of functional importance. Functions of testosterone include (1) development of secondary sex characteristics; (2) maintenance of the male duct system; (3) expression of male sexual behavior (libido); (4) function of the accessory glands; (5) function of the tunica dartos muscle in the scrotum; and (6) spermatocytogenesis. Reciprocal action of androgens with the gonadotrophins is necessary for control of normal reproductive function (Figure 4-5). Inhibin is important in this control because of its inhibition of the release of FSH from the anterior pituitary.

**4-4**
**Primary**
**reproductive**
**hormones of**
**the adrenal**
**cortex**

The adrenal cortex produces two classes of steroid hormones which have been associated with mineral metabolsim (mineralocorticoid) and carbohydrate metabolism (glucocorticoids). Glucocorticoids, the principal one being *cortisol*, have been classified as anti-stress hormones, also. While progestins, estrogens and androgens have been isolated from the adrenal cortex, they have not been seen in quantities high enough to affect the reproductive

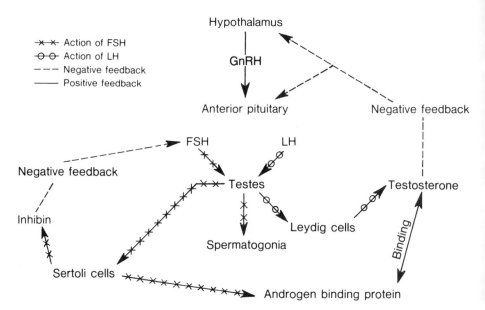

FIGURE 4-5. Interrelationship of the hormones regulating reproduction in the male.

processes. Some think that they may be released at levels high enough to alter normal reproduction during periods of severe stress, but verification of this has been difficult.

Recently, a role for glucocorticoids in the initiation of parturition in sheep has been demonstrated. Furthermore, it appears that the glucocorticoids involved in this process have been of fetal rather than maternal origin. This phenomenon has not been as clearly demonstrated in other classes of farm animals, but the evidence appears sufficiently strong to include glucocorticoids as a primary hormone of reproduction. In addition, a role for glucocorticoids in milk synthesis has been advanced (Chapter 10).

**4-5
Endocrine
function of the
uterine/placental
unit**

The placenta does not fit the classical definition of an endocrine gland but does assume an endocrine function during pregnancy. Estrogens, progestins, and relaxin are produced by the placenta in certain species and supplement production of these hormones by the ovaries. In addition, placental hormone(s) with luteotrophic and/or lactogenic activity have been identified in some species and may be present in others. *Human chorionic gonadotrophin* (HCG) has been isolated from the urine of pregnant women. Its principal action is LH-like and is believed to help maintain the function of the corpus luteum during pregnancy. *Pregnant mare serum gonadotrophin* (PMSG) is produced by endometrial cups which form when specialized cells in the chorion invade the endometrium of the pregnant uterus of the mare.

Principally, PMSG has FSH-like action. It has been isolated from the serum of mares during early pregnancy. A role for PMSG in the formation of accessory corpora lutea in the mare during pregnancy has been postulated. Both HCG and PMSG are proteins.

**4-6**
**Reproductive**
**role of**
**prostaglandins**

Prostaglandins are a group of biologically active lipids which have arachidonic acid, a 20-carbon, unsaturated fatty acid as their principal pre-curser. Based on differences in chemical structure, they are classified into two broad groups, prostaglandin E compounds (PGE) and prostaglandin F compounds (PGF). Prostaglandins have hormone-like actions but are not produced by a specific gland or tissue. Rather they are produced by cells throughout the body, including cells in the uterus and cells in the vesicular glands of the male. Although there are many different compounds with varying physiological activity in each group, $PGF_2\alpha$ has received the most attention in reproductive physiology.

$PGF_2\alpha$ will cause regression of the corpus luteum and has a stimulating effect on smooth muscles. Because of these actions, natural functions in the control of the estrous cycle, ovum transport, sperm transport, and parturi-tion have been proposed. Likewise, $PGF_2\alpha$ has been used in clinical situa-tions where regression of the corpus luteum or stimulation of smooth mus-cles is desired.

$PGE_2$ also appears to have an important role in regulation of reproduc-tive functions. Its effects are generally opposite to that of $PGF_2\alpha$. Notable is the anti-luteolytic effect on the corpus luteum. Thus, $PGE_2$ may serve to prevent the luteolytic effect of $PGF_2\alpha$.

Little research has been reported on a role for prostaglandins in the regulation of reproductive function in males. It has been demonstrated in bulls that injection of $PGF_2\alpha$ will cause a surge in LH and testosterone. However, an integrative role for prostaglandins in the natural regulation of reproductive function has not been determined.

**4-7**
**The pineal**
**gland**

The *pineal gland,* located above the hypothalamus, is considered by some to be an endocrine gland, but there is little knowledge of its function in farm animals. In studies with rats and hamsters, *melatonin* has been isolated from the pineal. Melatonin inhibits the function of the gonads. Given the effect of melatonin on the gonads and the sensitivity of the pineal gland to light, it has been suggested that this gland plays a role in the control of breeding season in seasonal breeders such as the ewe and mare.

**4-8**
**Regulation of**
**hormonal**
**receptor sites**

Hormone action is dependent on release of the hormone in question from its gland, transport to the target cells via the circulatory system and binding of the hormone to cellular receptor sites. After the hormone binds to the cellular receptor site, reactions are initiated within the cell to carry out the physiological response associated with the hormone.

The concentration of receptor sites for a specific hormone in a particular organ are dependent on the endocrine status of the animal. While research in this area is relatively new, and limited mostly to laboratory animals, some information is available on regulation of hormonal receptor sites. It provides a new basis for understanding how certain hormones synergize in regulating a physiological function. Patterns of regulation that can be seen are (1) hormones which regulate their own receptor sites; (2) synergism of two hormones to regulate the receptor sites of one of the hormones; and (3) hormones which regulate receptors of other hormones.

FSH increases its own receptor sites in the ovarian follicle and this action is speeded as estrogen levels increase. With estradiol priming, FSH also increases follicular receptors for LH. Estradiol increases its own receptors. LH causes a reduction in FSH, LH and estradiol receptors, but increases receptors for prolactin. In the forming corpus luteum, prolactin increases LH receptors and prevents LH-induced loss of LH receptors.

A similar pattern can be postulated for the testes. Prolactin will maintain the level of LH receptors in the cells of Leydig, preventing LH-induced losses. FSH receptors are found in both spermatogonia and Sertoli cells, indicating two probable sites of action for FSH.

In the uterus, estrogen increases the concentration of both estrogen and progesterone receptors. Progesterone blocks synthesis of new estrogen receptors, resulting in a reduction in their concentration.

Both estrogen and progesterone receptors are found in the hypothalamus and pituitary gland. This lends support to the hypothesis that these steroids exert feedback control on both the hypothalamus and anterior pituitary.

**4-9**
**Summary**

Most of the hormonal regulation of the reproductive processes is contained in the hypothalamic-anterior pituitary-gonadal axis. Releasing hormones from the hypothalamus controls the function of the anterior pituitary. Gonadotrophic hormones from the anterior pituitary control the function of the gonads, both in the production of gametes and hormones. In turn, through feedback mechanisms involving the hypothalamus, the steroid hormones and proteins of the gonads regulate the release of gonadotrophins. These gonadal steroids also maintain optimum conditions for fertility through their effects on mating behavior and maintenance of the female and male duct systems. While other hormones have important regulatory functions in reproduction, their function is dependent on the reciprocal balance between the gonadotrophic and steroid sex hormones. A greater appreciation for and understanding of the intricate balance needed for successful reproduction should evolve as the student progresses in his or her study of reproduction.

**Suggested reading**

Baird, D. T. 1972. Reproductive hormones. *Reproduction in Mammals. 3. Hormones in Reproduction.* ed. C. R. Austin and R. V. Short. Cambridge University Press.

Barraclough, C. A., P. M. Wise, J. Turgeon, D. Shander, L. DePaulo and N. Rance. 1979. Recent studies on the regulation of pituitary LH and FSH secretion. *Biol. Reprod.*, 20:86.

Bartke, A., A. A. Hafiez, F. J. Bex and S. Dalterio. 1978. Hormonal interactions in regulation of androgen secretion. *Biol. Reprod.*, 18:44.

Cross, B. A. 1972. The hypothalamus. *Reproduction in Mammals. 3. Hormones in Reproduction.* ed. C. R. Austin and R. V. Short. Cambridge University Press.

Erickson, G. F. and A. J. W. Hsueh. 1978. Secretion of "inhibin" by rat granulosa cells *in vitro. Endocrinology*, 103:1960.

Greep, R. O. ed. 1977. *Reproduction Physiology II. International Review of Physiology.* Vol. 13. University Park Press.

Koligan, K. B. and F. Stormshak. 1977. Nuclear cytoplasmic estrogen receptors in the ovine endometrium during the estrous cycle. *Endocrinology*, 101:524.

Kotite, N. J., S. N. Nayfeh and F. S. French. 1978. FSH and androgen regulation of Sertoli cell function in the immature rat. *Biol. Reprod.*, 18:65.

Nalbandov, A. V. 1976. The endocrinology of reproduction. Chapter 3, *Reproductive Physiology of Mammals and Birds.* W. H. Freeman and Co.

Niswender, G. D., T. M. Nett and A. M. Akbar. 1974. The hormones of reproduction. *Reproduction in Farm Animals.* (3rd ed.) ed. E. S. E. Hafez. Lea and Febiger.

Richards, J. S., J. I. Ireland, M. C. Rao, G. A. Bernarth, A. R. Midgely, Jr. and L. E. Reichert, Jr. 1976. Ovarian follicular development in the rat: Hormone receptor regulation by estradiol, follicle-stimulating hormone and luteinizing hormone. *Endocrinology*, 99:1562.

Richards, J. S. and J. J. Williams. 1976. Luteal cell receptor content for prolactin (PRL) and luteinizing hormone (LH): Regulation by LH and PRL. *Endocrinology*, 99:1571.

Schally, A. V. 1978. Aspects of hypothalamic regulation of the pituitary gland. *Science*, 202:18.

Schwabe, C., B. Steinetz, G. Weiss, A. Segaloff, J. K. Donald, E. O'Byrne, J. Hockman, B. Carriere and L. Goldsmith. 1978. Relaxin. *Rec. Prog. Hormone Res.*, 34:23.

Steinberger, A. and E. Steinberger. 1976. Secretion of an FSH-inhibiting factor by cultured Sertoli cells. *Endocrinology*, 99:918.

Tindall, D. J., C. R. Mena and A. R. Mena. 1978. Hormonal regulation of androgen binding protein in hypophysectomized rats. *Endocrinology*, 103:589.

Zipf, W. B., A. H. Payne and R. Kelch. 1978. Prolactin, growth hormone and luteinizing hormone in the maintenance of testicular luteinizing hormone. *Endocrinology*, 103:595.

Part

2

# Reproductive processes

# Chapter

# 5

# The estrous cycle

The estrous cycle is defined as the time between periods of estrus. The average length of the estrous cycle is similar for all farm species, albeit shorter for the ewe (Table 5-1). It is about 17 days for the ewe, 21 days for the cow, 22 days for the mare, and 20 days for the sow. Individual variation is seen in all species. Estrous cycles ranging from 17 to 24 days are considered normal in the cow and a range of 19 to 25 days is reported in the mare. While variation among individuals of a particular species is expected, variable cycles for one individual may indicate an abnormality.

**5-1
Puberty**

*Puberty* in females is defined as the age at the first expressed estrus with ovulation. It should not be considered sexual maturity. If animals are bred at puberty, a high percentage will have difficulty with parturition (Table 22-2). Since the ovaries will respond to exogenous gonadotrophins several months before puberty, it seems likely that puberty occurs when gonadotrophins are produced by the anterior pituitary in concentrations high enough to initiate follicle growth and ovulation. Follicle growth can be detected several months before puberty.

Age and weight at puberty are affected by genetic factors. This can be seen by comparing species or by comparing breeds within a species. Average age at puberty is 4 to 7 months for sows, 7 to 10 months for ewes, 8 to 11 months for European cows, and 15 to 24 months for mares. Weight at puberty for breeds within a given species depends on the mature size of the breed in question (Table 5-2). Jerseys reach puberty at about 8 months and 160 kg while for Holsteins 11 months and 270 kg is the average.

A number of environmental factors have a pronounced effect on age at puberty. In general, any factor which slows growth rate, thus preventing expression of full genetic potential, will delay puberty. A Holstein heifer on a recommended plane of nutrition will reach puberty at about 11 months of age, but if raised from birth on 62% of the recommended level of energy she

TABLE 5-1.   Species differences in various characteristics of the estrous cycle

|  | Cow | Ewe | Sow | Mare | Goat (Doe) |
|---|---|---|---|---|---|
| Estrous cycle (days) | 21 | 17 | 20 | 22 | 20 |
| Metestrus (days) | 3-4 | 2-3 | 2-3 | 2-3 | — |
| Diestrus (days) | 10-14 | 10-12 | 11-13 | 10-12 | — |
| Proestrus (days) | 3-4 | 2-3 | 3-4 | 2-3 | 2-3 |
| Estrus | 12-18 hrs | 24-36 hrs | 48-72 hrs | 4-8 days | 34-38 hrs |
| Ovulation | 10-12 hrs after estrus | late estrus | mid estrus | 1-2 days before estrus ends | late estrus |

TABLE 5-2.   Species and breed differences in age and weight at puberty

|  | Age mo | Weight kg |
|---|---|---|
| Sow | 4-7 | 68-90 |
| Ewe | 7-10 | 27-34 |
| Mare | 15-24 | (Varies with mature size of breed) |
| Cow | 8-13 | 160-270 |
| Jersey | 8 | 160 |
| Guernsey | 11 | 200 |
| Holstein | 11 | 270 |
| Ayrshire | 13 | 240 |

will be over 20 months of age at puberty (see Table 22-2). High environmental temperature delays puberty. Beef heifers, when reared at 10°C, reached puberty at 10.5 months of age, but similar heifers reared at 27°C were over 13 months of age at puberty. Gilts farrowed in the late spring reached puberty at a later age than gilts farrowed in other seasons because growth before puberty was slowed by hot, summer temperatures. Other environmental factors that might delay puberty include poor health and poor sanitation in rearing facilities. While adverse environments delay puberty and reduce the mature size of animals, weight at puberty is not greatly affected. Heifers on a low plane of nutrition were 84% older but only 7% smaller at puberty than well fed heifers. Feeding above recommended levels will result in earlier puberty. Holstein heifers fed at 146% of the recommended level reached puberty at an average of 9.2 months of age as compared to 11 months for controls receiving the recommended diet. Both problems associated with overconditioning and the extra cost of such a diet made overfeeding undesirable.

**5-2**
**Periods of the**
**estrous cycle**

The periods of the estrous cycle are *estrus, metestrus, diestrus,* and *proestrus* (Table 5-3). These periods occur in a cyclic and sequential manner, except for periods of *anestrus* (absence of cycling) in seasonal breeders such as the ewe and mare, as well as anestrus during pregnancy and the early postpartum period for all species.

TABLE 5-3.   Primary characteristics of the periods of the estrous cycle in the cow

| Period | Day(s) | Principal features |
|---|---|---|
| Estrus | 1 | Behavioral signs of estrus |
| Metestrus | 2-4 | Ovulation |
| | | Corpus luteum formation |
| Diestrus | 5-16 | Corpus luteum function |
| Proestrus | 17-21 | Rapid follicle growth |

**5-2.1**
**Estrus**

Estrus is defined as the period of time when the female is receptive to the male and will stand for mating. (See Section 5-4 for expanded discussion.) The length of the period of estrus varies between species (Table 5-1). Estrus lasts for 12 to 18 hours in the cow. As in the estrous cycle, considerable variation is seen between individuals. Also, cows in hot environments have shorter periods of estrus (10 to 12 hrs) than the average 18-hour period for cows in cool climates. Estrus lasts for 24 to 36 hours in the ewe, 40 to 72 hours in the sow and 4 to 8 days in the mare. The mare is the most variable of the farm species, with reported estrus ranging from 2 to 12 days. Ovulation is associated with estrus occurring 10 to 12 hours after the end of estrus in the cow, middle to late estrus in the ewe, about mid-estrus in the sow and 1 to 2 days before the end of estrus in the mare. The day of estrus is the first day of the estrous cycle in the cow. For other species the first day of estrus is the first day of the cycle.

**5-2.2**
**Metestrus**

The period of metestrus begins with the cessation of estrus and lasts for about 3 days. Primarily, it is a period of formation of the corpus luteum (corpora lutea with multiple ovulation). However, ovulation occurs during this period in cows. Also, a phenomenon known as metestrus bleeding occurs in cows, appearing in about about 90% of all metestrus periods in heifers and 45% in mature cows. During late proestrus and estrus, high estrogen concentrations increase the vascularity of the endometrium, this vascularity reaching its peak about 1 day after the end of estrus. With declining estrogen levels, some breakage of capillaries may occur resulting in a small loss of blood. This will be noticed as a patch of blood on the tail approximately 35 to 45 hours after the end of estrus. It is not an indication of conception or of a failure to conceive. Also, it should not be confused with menstrual bleeding which occurs in humans.

**5-2.3**
**Diestrus**

Diestrus is characterized as the period in the cycle when the corpus luteum is fully functional. In the cow it starts about day 5 of the cycle, when an increase in blood concentration of progesterone can first be detected, and ends with regression of the corpus luteum on day 16 or 17. For the sow and ewe it extends from about day 4 through days 13, 14, or 15. Mares are more variable because of the irregular length of estrus. For mares ovulating on day 5, diestrus will extend from approximately day 8 through day 19 or 20. It has been called the period of preparation of the uterus for pregnancy.

**5-2.4**
**Proestrus**

Proestrus begins with the regression of the corpus luteum and drop in progesterone and extends to the start of estrus. The principal distinguishing feature of proestrus is the occurrence of rapid follicle growth. Late during this period the effects of estrogen on the duct system and behavioral symptoms of approaching estrus can be observed.

**5-3**
**Hormonal**
**control of the**
**estrous cycle**

The estrous cycle is principally regulated through a reciprocal balance between the steroid hormones of the ovary and the protein, gonadotropic hormones of the anterior pituitary. While function of the anterior pituitary is controlled by the hypothalamus, the exact nature of this control has not been defined. Likewise, the precise mechanism for the release of $PGF_{2\alpha}$ from the uterus is not clear. $PGF_{2\alpha}$ is believed to cause regression of the corpus luteum during the estrous cycle (See Chapter 2).

By the use of radioimmunoassay techniques, supplemented by other chemical and bioassay studies, the relative changes in the blood levels of ovarian steroids and gonadotropic hormones have been monitored through the estrous cycle (Figures 5-1, 5-2, 5-3, 5-4). Similarities are more marked than differences among species. From this information, the conclusion that progesterone has a dominant effect on regulation of the estrous cycle seems quite logical. During diestrus, when progesterone concentrations are high, concentrations of FSH, LH, and total estrogens remain relatively low. Some follicle growth can be detected, and is associated with small mid-diestrus surges of FSH, LH, and estrogen in some species, but not the rapid follicle growth that is seen in the 2 to 3 day period before ovulation. Likewise during pregnancy, a high concentration of progesterone suppresses release of the gonadotrophic hormones that would initiate estrual behavior. Thus, progesterone regulates the release of gonadotrophins through a negative feedback control.

At the end of diestrus, $PGF_{2\alpha}$ from the uterus causes the corpus luteum to regress along with a sharp drop in the blood concentration of progesterone. This low progesterone concentration may serve as a stimulus or may remove a block from either the hypothalamus or anterior pituitary which results in a release of FSH, LH, and prolactin. A surge of estrogen seen during proestrus in the sow, ewe, and cow and near the start of estrus in the mare drops near the end of estrus. Surges of FSH and LH lasting 8 to 10

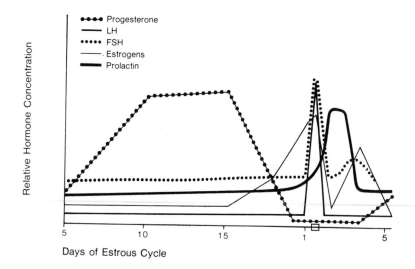

FIGURE 5-1. Hormonal changes in the peripheral plasma during the estrous cycle of the cow. The drop in progesterone on day 16, 17, or 18 is followed by surges in estrogens during late proestrus, FSH and LH during estrus, and prolactin during late estrus and early metestrus. (Based on literature.)

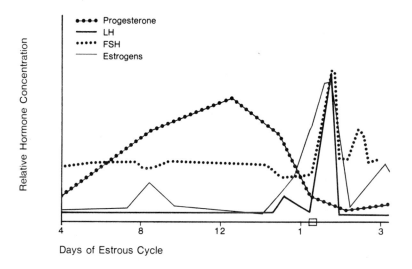

FIGURE 5-2. Hormonal changes in the peripheral plasma during the estrous cycle of the ewe. Patterns for the ewe are similar to that for other species. A reduction in FSH during proestrus is followed by a spike during estrus and another surge during metestrus. (Based on literature.)

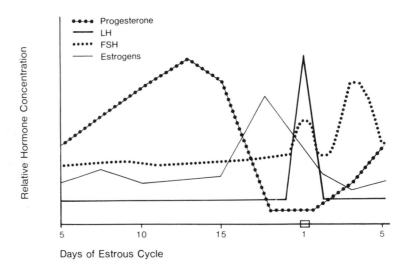

FIGURE 5-3. Hormonal changes in the peripheral plasma during the estrous cycle in the sow. Notable is the marked increase in FSH during proestrus. (Based on literature.)

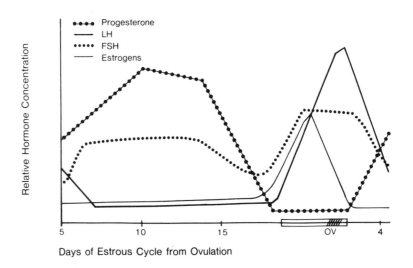

FIGURE 5-4. Hormonal changes during the estrous cycle in the mare. Patterns are similar to that of other species except that surges of FSH and LH during estrus last for several days. (Based on literature).

hours are seen during estrus within 24 hours preceding ovulation. Prolactin surges near the start of estrus and remains high through estrus. Since FSH stimulates the follicle growth that starts during proestrus and directly or indirectly the surge of estrogen, failure to detect a surge of FSH during proestrus cannot be completely explained at this time. Possibly a release of FSH from the anterior pituitary is masked by increased binding of FSH by receptor sites in the granulosa cells. Such a hypothesis is supported by some earlier research in cows indicating a depletion of FSH from the anterior pituitary about 2 days before the LH surge. The rapid surge of FSH and LH during estrus appears to be necessary for the final growth and maturation of the Graafian follicle that is necessary for ovulation. The functional significance of increased FSH during metestrus has not been determined.

With follicle growth, a surge in estrogen occurs. Estrogen is produced by cells in the Graafian follicle. The theoretical mode of synthesis described in section 3-1 involves synthesis of testosterone by the theca interna cells and its subsequent conversion to estrogen by the granulosa cells. While still in question, support for this theory is provided through research demonstrating that granulosa cells have receptor sites for testosterone and that binding of FSH to receptor sites in the ovary is limited to granulosa cells. A limitation to the theory is the question of whether testosterone can diffuse across the basement membrane, separating the theca interna cells and its capillary bed from the granulosa, rapidly enough to permit the measured surge of estrogen. Estrogen causes the behavioral and physiological signs of estrus and triggers the release of LH through a positive feedback control.

Following ovulation a corpus luteum will form at each ovulation site. Formation occurs rapidly and by day 4 or 5 of the estrous cycle a detectable increase in progesterone will again indicate diestrus. LH has the dominant controlling influence on formation of the corpus luteum. Prolactin synergizes with LH by increasing and maintaining hormone receptor sites for LH in the corpus luteum in at least some species. LH is luteotrophic and with prolactin maintains the function of the corpus luteum in farm animals. LH appears to maintain this function by greatly increasing the blood flow through the corpus luteum. Conversely, $PGF_{2\alpha}$ shuts off the blood flow to the corpus luteum depriving it of metabolites needed for synthesis of progestins, and causing its regression.

**5-4
Mating
behavior**

Estrus has been defined as the period when the female is receptive to the male. For controlled natural mating or for use of artificial insemination it is necessary that animal managers be able to recognize the signs of estrus and be aware of factors which contribute to normal estrual behavior.

High levels of estrogens have been associated with the behavioral signs of estrus and are of paramount importance. However, evidence exists for the interaction of estrogens with certain senses in eliciting the full behavioral response. Evidence for this is stronger in sows and ewes than in cows or mares, but the contribution of the senses is probably important in all species. Both the sense of smell and the sense of hearing are of demonstrated importance in the sow. When boars are not present, providing the sound of the boar through recordings, or the smell of the boar through solutions containing *pheromones* from the boar, have elicited a stronger estrual response. Pheromones are odorous chemical substances produced by one sex which attracts the opposite sex. While the ewe does not exhibit signs of estrus when the ram is not present, the senses related to this response have not been identified. A combination of the senses of sight, hearing, and smell may be involved. The sense of touch is probably important to the mating response of all species in that bunting, biting, licking, and rubbing are a part of the courtship before copulation.

In general, females will be more restless, irritable, and excitable during estrus. In addition, interest in the male will become apparent if the male of the species is in the vicinity. Such indications can first be seen during late proestrus, but the female will not stand to be mounted by the male or by another female during proestrus. In recognizing the signs of estrus or approaching estrus, knowledge of the personalities of the animals in question will be beneficial. Also, it is helpful to watch for females in estrus when they are quiet. Dawn and dusk are good times to watch for estrus, especially if the animals do not know they are being watched. If the females are excited by the presence of people, noises, or anticipation of feeding, detection of estrus is very difficult. Specific behavioral and physiological patterns are characteristic of the different species.

Cows are unique in that they display rather strong homosexual tendencies, making estrus detection comparatively easy even when bulls are not present. Cows in estrus will solicit mounts and attempt to mount other cows. They may smell the vulva of other cows. Frequently, they raise and switch their tail and may leave the herd in search of a bull. They will have a congested vulva and clear mucus can often be seen streaming from the vulva. Cows in other periods of the estrous cycle will mount cows in the period of estrus, but will not stand to be mounted. Therefore, standing for mounting is the strongest single behavioral indication of estrus.

In contrast to the cow, the ewe displays no signs of estrus if the ram is not present. The ewe will rub the neck and body of the ram. She will roam around the ram smelling his genitalia and shaking her tail vigorously. The vulva of the ewe will not be congested, and there will be no visible mucus. If using artificial insemination in ewes, use of altered rams is necessary for detection of estrus. (See Chapter 18 for discussion of altered males.)

Sows will assume a mating stance when pressure is applied to their rump by a boar, another sow, or the hand of an attendant. This provides some convenience in artificial insemination in that sows can be inseminated without restraint if pressure is maintained on the rump. There will be no visible mucus during estrus, but the vulva will be swollen and congested.

The swollen vulva is more noticeable in gilts than sows. The vulva may become swollen after giving certain medications, so this should be considered with other signs of estrus.

The mare will allow the stallion to smell and bite. She will extend her hind legs, lift her tail to the side and lower her rump. The vulva will be elongated and swollen, with the labia partly everted. The erect clitoris will be exposed frequently by contractions (winking) of the labia. The mare should be teased by a stallion for accurate detection. Any attempt to fight the stallion indicates she is not in estrus even though some other signs of estrus are apparent. In mares and other species knowledge of their individual behavior during estrus will aid in detecting estrus.

Accurate estrus detection has been listed as a major reproductive problem in farm animals. Paramount to accurate detection of estrus is an understanding of the expected behavior for the species in question and the factors which contribute to the expected response. Additional discussion on detection of estrus is found in Chapter 19.

**5-5**
**Seasonal**
**breeders**

Most wild species have a breeding season that is initiated at a time when the environment will allow for the best survival of the young at their birth. These patterns have developed through natural selection over many generations. Patterns of seasonal breeding range from species that have one period of estrus each year (monoestrus) to species that have a series of estrous cycles limited to a portion of the year (seasonally polyestrus). All domesticated animals probably exhibited seasonal breeding tendencies before domestication. This has been changed by providing better environments (housing and nutrition) and by selecting for more prolific animals. Seasonal changes in fertility in cattle and swine can be related to adverse environmental conditions that are present in some years but not in others. True seasonal breeding patterns are inherent in ewes and mares.

**5-5.1**
**Ewes**

Most breeds of sheep exhibit seasonal breeding patterns. Sheep native to the tropics are an exception, and will cycle throughout the year. While breed differences are apparent, there is a tendency for sheep native to arctic regions to have shorter seasons than those found in temperate regions.

Sheep are short-day or fall breeders (Figure 5-5). Their breeding season is initiated as the ratio of daylight to darkness decreases and ends when increasing daylengths reach a ratio of nearly equal daylight and darkness. For most breeds the season falls between the autumnal equinox and spring equinox. However, for the Dorset Horn the season starts in late June and extends to late April. Silent estrous periods (ovulation without behavioral estrus) occur at the beginning and at the end of the breeding seasons for all breeds.

The daylength pattern has a dominant controlling influence on initiation and termination of breeding season. If sheep are shipped from the northern

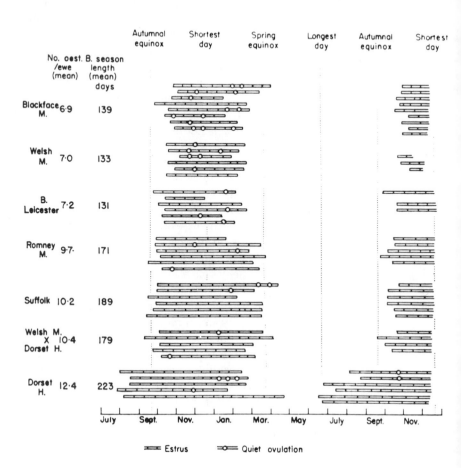

FIGURE 5-5.  Breed differences in the duration of the breeding seasons in adult ewes in Great Britain. (Hafez. J. Agri. Sci. 42:305. 1952.)

hemisphere to the southern hemisphere, the breeding season will reverse. A similar reversal can be achieved by controlled artificial lighting in a room that excludes all natural light. Altering the temperature in environmental control chambers does not influence breeding season unless light patterns are altered also.

5-5.2
Mares

Mares are long-day breeders. Their season is initiated as the ratio of daylight to darkness increases and ends during decreasing daylengths. The average season for ponies is May to October, but is longer in horses, extending from February to November. Behavioral estrus sometimes occurs in the short-day months, but is not accompanied by ovulation. Much variation in the length of the breeding season is seen in individual mares and among mares. As in the ewe, the daylength pattern has the dominant controlling influence on the mare's breeding season. The breeding season can be lengthened by the use of artificial light to lengthen the light period before the start or near the end of the anticipated season.

In both the ewe and mare, it is theorized that the light stimulus sends signals along the optic nerve to the hypothalamus, which then initiates release of gonadotrophins to start the estrous cycle. A role for the pineal gland has been advanced, but the exact relationship is not clear. Denervation of the pineal gland will delay the onset of the breeding season in mares.

**Suggested reading**

Dobson, H. 1978. Plasma gonadotrophins and oestradiol during oestrus in the cos. *J. Reprod. Fert.*, 52:51.

Dziuk, P. H. 1977. Reproduction in pigs. *Reproduction in Domestic Animals.* ed. H. H. Cole and P. T. Cupps. Academic Press.

Garverick, H. A., R. E. Erb, G. D. Niswender and C. J. Callahan. 1971. Reproductive steroids in the bovine. III. Changes during the estrous cycle. *J. Anim. Sci.*, 32:946.

Ginther, O. J. 1979. Reproductive seasonality in mares. *Beltsville Symposia in Agricultural Research 3. Animal Reproduction.* ed. H. W. Hawk. Allanheld, Osmun and Co. Publishers; Halsted Press, a division of Wiley & Sons.

Hafs, H. D. and M. S. McCarthy. 1979. Endocrine control of testicular function. *Beltsville Symposia in Agricultural Research. 3. Animal Reproduction.* ed. H. W. Hawk. Allanheld, Osmun and Co. Publishers; Halsted Press, a division of Wiley & Sons.

Hansel, W., P. W. Concannon and J. H. Lukaszewska. 1973. Corpora lutea of large domestic animals. *Biol. Reprod.*, 8:222.

Hansel, W. and E. Ecternkamp. 1972. Control of ovarian function in domestic animals. *Amer. Zool.*, 12:225.

Hansel, W. and R. B. Snook. 1970. Pituitary ovarian relationships in the cow. *J. Dairy Sci.*, 53:945.

L'Hermite, M. G. D. Niswender, L. E. Reichert, Jr., and A. R. Midgeley, Jr. 1972. Serum follicle-stimulating hormone as measured by radioimmunoassay. *Biol. Reprod.*, 6:325.

Rayford, P. L., H. J. Brinkley, E. P. Young and L. E. Reichert, Jr. 1974. Radioimmunoassay of porcine FSH. *J. Anim. Sci.*, 39:348.

Robertson, H. A. 1977. Reproduction in the ewe and goat. *Reproduction in Domestic Animals.* (3rd. ed.) ed. H. H. Cole and P. T. Cupps. Academic Press.

Robinson, T. J. 1977. Reproduction in cattle. *Reproduction in Domestic Animals.* (3rd. ed.) ed. H. H. Cole and P. T. Cupps. Academic Press.

Short, R. V. 1972. Role of hormones in sex cycles. *Reproduction in Mammals. 3. Hormones of Reproduction.* ed. C. R. Austin and R. V. Short. Cambridge University Press.

Stabenfeldt, G. H. and J. P. Hughes. 1977. Reproduction in horses. *Reproduction in Domestic Animals.* (3rd. ed.) ed. H. H. Cole and P. T. Cupps. Academic Press.

# 6

# Spermatogenesis and maturation of spermatozoa

*Spermatogenesis* is the process by which spermatozoa are formed. This process occurs in the seminiferous tubules (Chapter 3). Output of spermatozoa per day has been reported to be 4 billion for beef bulls, 7 billion for dairy bulls, 8 billion for rams, 10 billion for stallions, and 15 to 20 billion for boars. Actual production of spermatozoa may be 50 to 100% higher, because all that are produced cannot be collected. After formation in the seminiferous tubules, spermatozoa will be forced through the rete testis and vasa efferentia into the epididymis where they are stored while undergoing maturation changes that make them capable of fertilization. After puberty, spermatogenesis will proceed as a continuous process throughout the life of the male. Seasonal changes in spermatozoa production may occur due to ambient temperature in all species, and due to light in rams. Reciprocal action of FSH, LH, and testosterone is necessary for the maintenance of spermatogenesis.

**6-1**
**Puberty**

In the male, puberty can be defined less succinctly than in the female. Generally, it is considered the time when spermatogenesis starts. If defined as the time when fertile spermatozoa are in the ejaculate, the age will be 10 to 12 months for bulls, 4 to 6 months for rams, 4 to 8 months for boars, and 13 to 18 months for stallions. However, spermatozoa are formed in the seminiferous tubules several weeks before they are seen in the ejaculate. In bulls, the time from appearance in the seminiferous tubules to appearance in the ejaculate is approximately 10 weeks.

A number of other changes can be seen in males, starting several weeks before fertile spermatozoa are in the ejaculate (Figure 6-1). These include changes in body conformation, increased aggressiveness and sexual desire, rapid growth of the penis and testes, and separation of the penis from the prepuce so that extension of the penis is possible. Timing of these events varies with species.

(a) Mature stallion. Note heavy shoulder and thick neck.

(b) Six month old son of stallion.

(c) Mature Beefmaster bull. Note crest, heavy front quarters, and well developed dewlap.

(d) Six month old son of bull.

FIGURE 6-1.  Secondary sex characteristics that develop following puberty.

Development of testicular function is primary to the changes observed as puberty approaches. This development is regulated by the endocrine system. LH is necessary for the development of the Leydig cells and for their function. However, during the period around puberty synergistic effects from FSH and prolactin have been reported. FSH and prolactin appear to make the Leydig cells more responsive to LH in young males by increasing and maintaining receptor sites for LH. As the Leydig cells develop and become functional, increasing concentrations of testosterone will stimulate most other changes associated with approaching puberty (see Chapter 4 for functions of testosterone). Synergistic effects from testosterone and FSH stimulate development of Sertoli cells, production of androgen binding protein and preparation of the seminiferous tubules for production of spermatozoa.

As with the female, puberty is not sexual maturity in the male. Some rams and boars are used for breeding and are highly fertile after about 6 months of age. However, the testes size and total production of spermatozoa increases until about 18 months of age. In bulls and stallions total production of spermatozoa will increase at least to 3 years of age. It is recommended that stallions not be used heavily until they are 3 to 4 years old. There is a high correlation between the size of the testes and total spermatozoa production.

All factors which affect age at puberty in females will affect age at puberty in males (Chapter 5). Genetic effects on puberty are seen by comparing species or breeds within a species. Any adverse environmental factor which slows growth rate will delay puberty. For example, male lambs on a low plane of nutrition may not reach puberty until after 12 months of age.

**6-2**
**The process of**
**spermatogenesis**

Spermatogenesis can be divided into two distinct phases (Figure 6-2). The first is *spermatocytogenesis*, a series of divisions during which spermatogonia form spermatids. The second is *spermiogenesis*, a phase where spermatids undergo a metamorphosis forming spermatozoa. The entire process takes about 7 weeks (49 days) in rams, but appears to require an additional 7 to 10 days in bulls. As spermatogenesis proceeds, the developing gametes migrate from the basement membrane of the seminiferous tubules toward the lumen.

6-2.1
Spermatocyto-
genesis

As stated in Chapter 3, the seminiferous tubules contain two types of cells (Figure 6-3). Sertoli cells, the larger and less numerous cells, serve as nurse cells. They may play this sustaining role during both spermatocytogenesis and spermiogenesis. Spermatogonia, the smaller and more numerous cells, are the potential gametes.

The first step in spermatocytogenesis (and thus, spermatogenesis) is a mitotic division of a spermatogonium forming a *dormant* and an *active spermatogonium*. The dormant spermatogonium remains in the germinal epithelium near the basement membrane to repeat the process later. The active spermatogonium will undergo four mitotic divisions, eventually forming 16 *primary spermatocytes* (Figure 6-2). In rams, these mitotic divisions are completed in 15 to 17 days. During the next step, each primary spermatocyte will undergo a meiotic division forming two *secondary spermatocytes*. With this division, the chromosome complement in the nucleus is reduced by half so that nuclei in secondary spermatocytes contain unpaired (n) chromosomes. This step requires approximately 15 days. Within a few hours after their formation each secondary spermatocyte will again divide forming two *spermatids*. Thus, four spermatids form from each primary spermatocyte, or 64 from each active spermatogonium.

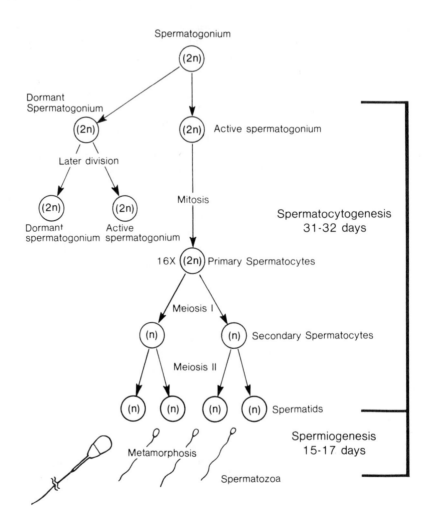

FIGURE 6-2. Spermatogenesis indicating the sequence of events and time involved in spermatogenesis in the ram. One spermatozoon is enlarged to permit more detail in morphology.

6-2.2
Spermiogenesis

During this phase spermatids are attached to Sertoli cells. Each spermatid undergoes a metamorphosis (change in morphology) forming a spermatozoon. During this metamorphosis the nuclear material will compact in one part of the cell forming the head of the spermatozoon, while the rest of the cell elongates forming the tail. The *acrosome,* a cap around the head of the spermatozoon, will form from the Golgi apparatus of the spermatid. As

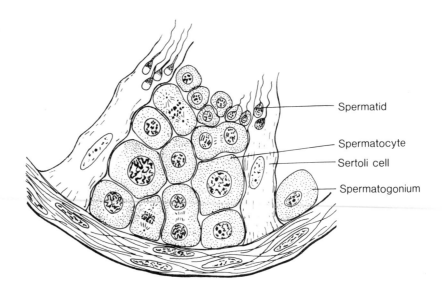

FIGURE 6-3. Small segment of an active seminiferous tubule showing the developmental stages that occur during spermatogenesis. Note the concentric layers of spermatogonia, spermatocytes, and spermatids progressing from the wall of the seminiferous tubules to the lumen. (Redrawn from Patten. 1964. *Foundation of Embryology.* (2nd ed.) McGraw-Hill.)

the cytoplasm from the spermatid is cast off during formation of the tail, a *cytoplasmic droplet* will form on the neck of the spermatozoon. Newly formed spermatozoa will then be released from the Sertoli cell and forced out through the lumen of the seminiferous tubules into the rete testis. Spermatozoa are unique cells in that they have no cytoplasm, and after maturation possess the ability to be progressively motile. Spermiogenesis is completed in 15 to 17 days.

**6-2.3
Hormonal
control of
spermatogenesis**

The endocrinology of reproduction has not been studied in males as extensively as in females. In bulls and rams there are 3 to 7 surges in LH per day followed by similar surges in testosterone (Figure 6-4). The principal role of LH in regulation of spermatogenesis appears to be indirect in that it stimulates the release of testosterone. Testosterone and FSH then act on the seminiferous tubules to stimulate spermatogenesis. Testosterone is necessary for certain steps in spermatocytogenesis and appears more dominant in the regulation of this process. On the other hand FSH appears more dominant in regulating spermiogenesis. Both testosterone and FSH appear to exert their influence directly through germ cells and/or indirectly through Sertoli cells. FSH stimulates the Sertoli cells to secrete both androgen binding protein (ABP) and inhibin. ABP may simply be a carrier for testos-

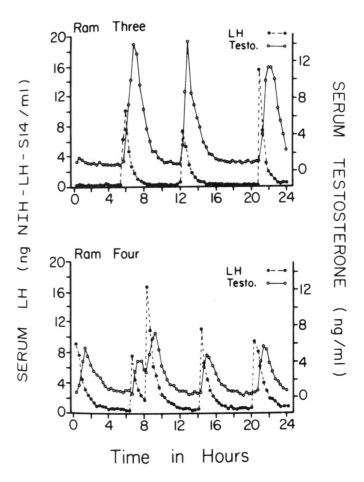

FIGURE 6-4.  Diurnal secretory pattern of LH and testosterone in mature rams. (Sanford, et al., 1974. *Endocrinology,* 95:627.) Similar patterns have been reported for the bull.

terone, making it more readily available during spermatogenesis in the seminiferous tubules and transporting it through the rete testis, vasa efferentia to the epididymis. ABP is absorbed in the epididymis. Feedback controls operating between the testis, hypothalamus, and anterior pituitary in regulating the release of gonadotrophins (FSH and LH) and gonadal steroids (testosterone) are probably similar to those described for the female (refer to Figures 4-4 and 4-5). It has been demonstrated that $PGF_{2\alpha}$ will stimulate the release of LH and testosterone. Therefore, $PGF_{2\alpha}$ may be involved in the feedback regulation between the hypothalamus, anterior pituitary, and testes.

6-2.4
Summation of
spermatogenesis

You have noted that 64 spermatozoa form from one active spermatogonium. When one considers that this type of activity is occurring continuously along the entire surface of the seminiferous tubules, it is easy to visualize how a mature ram can produce more than 8 billion spermatozoa in a day. Further, spermatogenesis is not the same type of cyclic process as is ovigenesis in the female. New spermatozoa are being formed and released to the duct system continuously.

The first step in spermatogenesis is important to the continuity of the process. The "seed" to make spermatogenesis continuous throughout the life of the male is the dormant spermatogonium. About one week after the active spermatogonium has gone through the divisions to form new spermatozoa, the dormant spermatogonium will divide forming a "new" active and a "new" dormant spermatogonium. The "new" active will proceed through spermatogenesis and the "new" dormant will later divide forming another active and another dormant, and so on. Without formation of a dormant spermatogonium, the supply of potential spermatozoa would diminish as does the number of potential ova in an ovary.

The process of spermatogenesis which has been described for the ram is probably similar in other species. Differences may exist as to the actual time involved in spermatogenesis. For example, some estimate that approximately 8 to 9 weeks are required for the completion of spermatogenesis in bulls. Differences may also exist as to the number of spermatozoa produced by a single active spermatogonium. However, the number of spermatozoa produced by each primary spermatocyte does not differ.

6-3
The
seminiferous
epithelial cycle
and
spermatogenic
wave

As stated in the previous section spermatogenesis does not proceed in the same cyclic manner as does ovigenesis in the female. While an active spermatogonium is going through the divisions necessary to form spermatozoa, several other spermatogonia in the same part of the seminiferous tubules will start spermatogenesis. Therefore, if a transverse section is cut from a seminiferous tubule, several generations of germ cells will be found. These are arranged concentrically with layers of spermatogonia near the wall of seminiferous tubules followed by spermatocytes and spermatids in layers progressing toward the lumen (Figure 6-3).

It has been determined that a kind of synchrony exists in the seminiferous tubules of a mature male in that certain cell types are always associated together. For example, a spermatid before the start of spermiogenesis is always associated with the same types of evolving spermatocytes and spermatogonia. Likewise, spermatozoa being released into the lumen have a unique association with specific types of evolving spermatocytes and spermatogonia. Based on the appearance of germ cells, eight different cellular associations have been identified in the seminiferous tubules of the male (Figure 6-5). For any given area of the seminiferous tubules these cellular associations occur in series and in cyclic regularity. The time between two successive appearances of the same cellular associations at a given point is the *seminiferous epithelial cycle*. The seminiferous epithelial cycle is similar in the bull and ram. It is also similar for the boar and stallion but they differ from the bull and ram.

The same eight cellular associations have been used to identify *spermatogenic waves* in the seminiferous tubules. These cellular associations occur in sequence along the seminiferous tubules, just as they do in one section of the tubules over a period of time. If the cellular association found at a specific point in a seminiferous tubule is identified as stage three, then stages two and four will be found on either side of this point. A complete

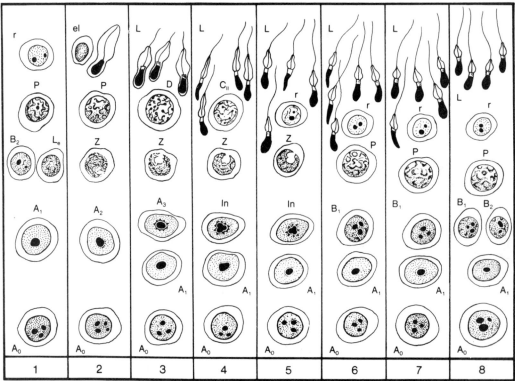

$A_0$, $A_1$, $A_2$, $A_3$, $B_1$, $B_2$, In = Spermatogonia; $L_e$, Z, P, D, $C_{II}$ = Spermatocytes; r, el, L = Spermatids

FIGURE 6-5. Cellular composition of the seminiferous epithelial cycle in the bull. These associations or stages are identified by the morphological changes of germ cell nuclei and local arrangements of spermatids. (Ortavant, 1977. *Reproduction in Domestic Animals.* (3rd ed.) ed. Cole and Cupps. Academic Press.)

series of the eight cellular associations along a seminiferous tubule is called a spermatogenic wave.

Knowledge of the seminiferous epithelial cycle and spermatogenic wave has potential utility in diagnosing certain abnormalities in males.

**6-4**
**Capacitation of spermatozoa**

After spermatozoa are produced in the seminiferous tubules two maturation processes are necessary before they can participate in fertilization. The first of these occurs in the epididymis, as discussed in Chapter 3. This was described as (1) gaining the ability to be motile, (2) gaining the ability to be fertile, and (3) losing the cytoplasmic droplet. Spermatozoa cannot penetrate the zona pellucida and participate in fertilization until they have undergone a second maturation process known as *capacitation*.

Capacitation is the final maturation of spermatozoa and occurs in the female tract. The process is not completely understood, but may involve removal of a lipoprotein layer from the surface of spermatozoa, thus permitting release of enzymes from spermatozoa that are necessary for penetration of the *zona pellucida* (outer membrane of the oocyte). In rabbits, it is known that spermatozoa must be in the uterus or oviduct for about four hours before one will finally stick to the zona pellucida and penetrate it.

There is indirect evidence that capacitation is necessary before fertilization in farm species. For example, best fertilization rates occur if cows are inseminated from middle to late estrus, some 12 to 18 hours before the estimated time of ovulation (Table 6-1). This is true even though spermatozoa will be in the oviduct near the site of fertilization several hours before the oocyte arrives. If capacitation was not needed, it seems likely that best fertilization rates would occur when insemination was timed to coincide with ovulation.

The oocyte may remain fertile for approximately 8 to 18 hours after ovulation (Table 7-2). For most species, spermatozoa are fertile for about 24 to 48 hours after deposition into the female tract. An exception is horses where spermatozoa in the tract of the mare are fertile for 2 to 6 days. While both male and female gametes remain fertile for several hours, an aging process begins soon after their release into the female tract. For example, best conception in dairy cows occurs when they are bred during late estrus as opposed to early estrus or near the time of ovulation. This relates both to the freshness of the gametes and to time needed for sperm capacitation. Even though the spermatozoa of the stallion may be fertile for 6 days, horse breeders report better results when the mare is bred every other day through estrus. Some horse breeders wait until a 35 mm Graafian follicle can be palpated before breeding, in an effort to synchronize insemination to about 24 hours preovulation. The best chance for fertilization and subsequent production of a viable embryo would be when the oocyte reaches the site of fertilization at a time when spermatozoa have been recently capacitated. For example, if an animal is inseminated as much as 24 hours

TABLE 6-1. Effect of time of insemination on ovulation and fertility in cows (cows normally ovulate 10 to 12 hours after the end of estrus)

| Time of breeding | Total cows | Cows conceiving from one service |
|---|---|---|
| Start of estrus | 25 | 44.0% |
| Middle of estrus | 40 | 82.5 |
| End of estrus | 40 | 75.0 |
| After estrus | | |
| 6 hours | 40 | 63.4 |
| 12 hours | 25 | 32.0 |
| 18 hours | 25 | 28.0 |
| 24 hours | 25 | 12.0 |
| 36 hours | 25 | 8.0 |
| 48 hours | 25 | 0.0 |
| Routine breeding | 194 | 63.4 |

Source: Trimberger and Davis, Univ. Neb. Res. Bul. 129, 1943.

before ovulation, spermatozoa may capacitate and start aging before ovulation. If inseminated at the time of ovulation, the ovum may be aging before spermatozoa capacitate. One should recognize that fertilization may occur even though one of the gametes is aged. With an aged gamete, the chances of early embryo loss increases (Section 7-5).

**Suggested reading**

Amann, R. P. 1970. Sperm production rates. *The Testis*. Vol. 1. ed. A. D. Johnson, W. R. Gomes and N. L. VanDemark. Academic Press.

Arey, L. B. 1947. *Developmental Anatomy*. (5th ed.) W. B. Saunders.

Bedford, J. N. 1970. Sperm capacitation and fertilization in mammals. *Biol. Reprod.*, 2:128.

Dym, M. 1977. The role of the Sertoli cell in spermatogenesis. *Reproductive Systems*. ed. R. Yates and M. Gordon. Raven Press.

Dym, M. and J. C. Cavicchia. 1978. Functional morphology of the testis. *Biol. Reprod.*, 18:1.

Gomes, W. R. 1978. Formation, migration, maturation and ejaculation of spermatozoa. *Physiology of Reproduction and Artificial Insemination of Cattle*. (2nd ed.) ed. G. W. Salisbury, N. L. VanDemark and J. R. Lodge. W. H. Freeman and Co.

Monesi, V. 1972. Spermatogenesis and spermatozoa. *Reproduction in Mammals. 1. Germ Cells and Fertilization*. ed. C. R. Austin and R. V. Short. Cambridge University Press.

Ortavant, R., M. Courot and M. T. Hochereau de Reviers. 1977. Spermatogenesis in domestic mammals. *Reproduction in Domestic Animals.* (3rd ed.) ed. H. H. Cole and P. T. Cupps. Academic Press.

Sanford, L. M., J. S. D. Winter, W. M. Palmer and B. E. Howland. 1974. The profile of LH and testosterone secretion in the ram. *Endocrinology,* 95:627.

Steinberger, E. 1971. Hormonal control of spermatogenesis. *Physiol. Rev.,* 51:1.

Young, W. C. 1931. A study of the function of the epididymis. III. Functional changes undergone by spermatozoa during their passage through the epididymis and vas deferens. *J. Expt. Biol.,* 8:151.

# Chapter

# 7

# Ovigenesis and fertilization

## 7-1
## Ovigenesis

Formation and maturation of the gametes must be completed for both the male and female before the reproductive processes can be initiated. *Ovigenesis* (or *oogenesis*) is the formation and maturation of the female gamete.

Ovigenesis (Figure 7-1) begins in the prenatal period. Formation of primary follicles has been described in Chapter 2. The potential gamete associated with the primary follicle when first formed is the oogonium. Oogonia originate from an extension of the yolk sac which forms from the hind gut of the embryo. Following initial formation, proliferation of oogonia by mitotic division occurs within the parenchyma. As previously stated in Chapter 2 this proliferation ceases before birth so that the ovaries, at birth, contain a fixed number of potential ova, or *oocytes*.

During the prenatal period and continuing postnatally a cyclic pattern in oocyte maturation has been reported. However, until the female reaches puberty no oocytes will reach full maturity. Those oocytes that start development before puberty become atretic and are lost as potential ova.

Maturation of oocytes will continue in a cyclic manner after puberty. During each estrous cycle a group of oocytes will start maturation while others remain dormant. Usually only one of the group that starts development will reach maturity and be released through ovulation to the duct system for possible fertilization in the cow, mare, and ewe. The other oocytes become atretic. In sows, 10 to 25 may reach maturity and be ovulated.

The first step in maturation involves growth (Figure 7-1). This includes an enlargement of the oocyte and proliferation of the follicular cells surrounding the oocyte. The *zona pellucida*, a gel-like outer membrane, forms around the oocyte.

After attaining its full size, the primary oocyte undergoes the first of two *meiotic divisions*. The products of the first meiotic division are the *secondary oocyte* and the *first polar body* which is trapped between the *vitelline mem-*

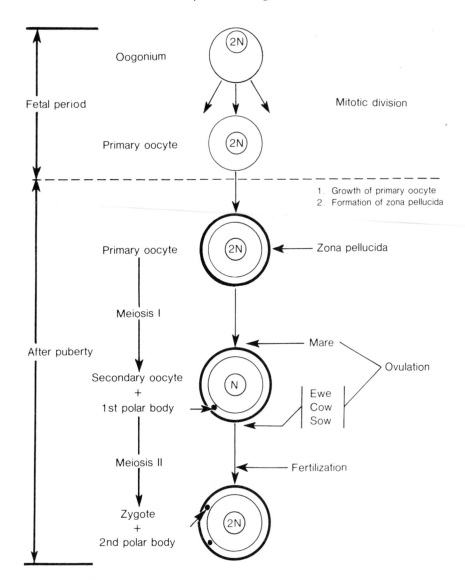

FIGURE 7-1. Principal maturation stages for the germ cell during ovigenesis.

*brane* and the zona pellucida in the perivitelline space. With this division, the chromosome numbers in the oocyte are changed from the *diploid* (2n) to the *haploid* (n) state. The secondary oocyte retains all of the cytoplasm and half of the nuclear material (chromosomes) of the primary oocyte. The other half of the nuclear material is extruded as the first polar body. This first meiotic division is completed just before ovulation in the cow, sow, and ewe and shortly after ovulation in the mare.

The second meiotic division begins immediately after completion of the first division. However, this division will not be completed unless fertilization occurs. With fertilization, the products of the second meiotic division are the *zygote* (fertilized egg) and the *second polar body*. It should be noted that the *true ovum* never exists in the cow, sow, ewe, or mare, or if so, only in a transient state. The true ovum would be the product of the second maturation division if the division was completed before fertilization.

**7-2**
**Ovulation and**
**gamete**
**transport**

**7-2.1**
**Oocyte**

With maturation of the oocyte and follicle, LH will stimulate ovulation. This involves rupture of the Graafian follicle. The contents of the follicle, which includes liquor folliculi, the secondary oocyte, and certain granulosa cells, will flow into the peritoneal cavity near the infundibulum. The oocyte will be imbedded in a sticky matrix containing granulosa cells (Figure 7-2). Both corona radiata and cumulus cells will be a part of this mass containing the oocyte. These granulosa cells are shed quickly in some species and not believed to be present at the time of fertilization. However, they appear to be a factor in the capture of the oocyte by the infundibulum and its movement into the ampulla. Most of the epithelial cells lining the infundibulum and ampulla are ciliated. These cilia beat in the direction of the uterus. The oocyte surrounded by its sticky mass of granulosa cells is readily picked up by these cilia. The directional flow of the oocyte is also aided by the flow currents of oviductal fluids. The directional flow of these currents is created by cilia beating in the direction of the uterus. Movement of the oocyte into and down the ampulla is aided by these currents. In addition to the current flow created by the action of the cilia, peristaltic contraction in the ampulla, which proceed in the direction of the uterus, tends to milk the oocyte down the ampulla. Therefore, both fluid flow created by the beating of cilia and

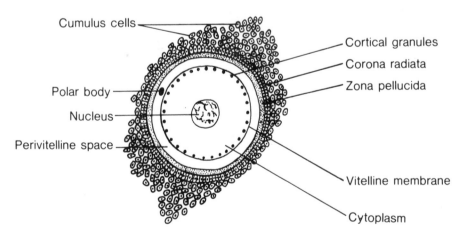

FIGURE 7-2. Oocyte and its associated cells soon after ovulation. Cumulus and corona radiata cells are shed before fertilization in some species.

peristaltic contractions contribute to oocyte transport. The oocyte passes through the ampulla to the ampullary-isthmic junction rapidly, then remains at that point for 2 to 3 days before moving down the isthmus to the uterus (Table 7-1). Thus, fertilization likely occurs at the ampullary-isthmic junction. Estrogens have been reported to cause retention of the oocyte in the oviduct, whereas progesterone hastens transport. Epinephrine has been reported to hasten transport also.

TABLE 7-1.   Transport time of ova in the oviduct of farm animals

| Species | Time-hrs |
| --- | --- |
| Cattle | 90 |
| Sheep | 72 |
| Horse | 98 |
| Swine | 50 |

Adapted from Hafez. *Reproduction in Farm Animals*. (3rd ed.), Lea and Febiger, 1974.

**7-2.2**
**Spermatozoa**

Mechanisms for transport of spermatozoa to the site of fertilization are more speculative than oocyte transport. Some spermatozoa reach the ampullary-isthmic junction within a few minutes after deposition of semen in the female tract. They reach the site of fertilization too quickly to have gotten there on the basis of their own motility. In addition, dead spermatozoa are transported to this site as quickly as motile sperm. Even though billions of spermatozoa are deposited in the female tract during natural service and millions through artificial insemination, only a few hundred actually reach the site of fertilization. Therefore, both barriers to transport and mechanisms which aid transport must be considered.

For most species, with deposit of semen in the vagina during natural service, the cervix is the greatest barrier to transport. The billions of spermatozoa deposited in the vagina are probably reduced to thousands actually reaching the uterus. The vagina is not a favorable environment for the survival of sperm, whereas the cervical canal, with its secretions at the time of estrus, does provide a most favorable environment. The cervical mucus is favorable for sperm because of its antibacterial properties and its slight alkalinity. The cervical mucus takes on a biophysical configuration during estrus that channels spermatozoa through the cervix into the uterus. It is likely that some spermatozoa are also temporarily trapped in the folds of the cervix with these spermatozoa being fed into the uterus slowly, thus, feeding fertile spermatozoa into the uterus for some hours after copulation.

When spermatozoa reach the body of the uterus via either natural transport through the cervix or artificial insemination, it is likely that uterine and oviductal contractions contribute to their transport. These contractions may be triggered by copulation with either neural transmitters (norepinephrine) or hormonal agents (oxytocin and/or $PGF_{2\alpha}$) interacting with estrogen to cause the contractions. There is no evidence that spermatozoa are preferentially routed toward the oviduct containing the oocyte. Therefore, approximately equal numbers will move up each oviduct. Random movement of spermatozoa may result in some movement back into the cervix also. Lack of preferential routing and random movement of spermatozoa likely account for the reduced number actually reaching the site of fertilization. The small size of the lumen through the uterotubal junction is another barrier for spermatozoa.

Both flow of fluids through the oviduct and contraction of smooth muscles in the walls of the isthmus aid transport of spermatozoa through the oviduct. Oviductal fluids flow in the direction of the ovaries at the uterine end of the oviduct. Action of cilia in the isthmus is believed to account for the directional flow of these fluids. Contractions in the walls of the isthmus, which also facilitate movement of spermatozoa to the site of fertilization, may be stimulated by either neural or hormonal agents.

## 7-3
## Fertilization

The process of fertilization starts with a collision between the oocyte and a spermatozoon and ends when their pronuclei have merged. The resulting diploid cell containing the genetic code for a new individual is the zygote. Even though the corona radiata and other granulosa cells may be shed before this process begins, fertilization will be described as though they remain around the oocyte. Features of the oocyte which are important to the process of fertilization are illustrated in Figure 7-2. These include the zona pellucida, the gel-like outer membrane of the oocyte; the vitelline membrane, which is the true plasma membrane of the oocyte; cortical granules, small rounded bodies which lie just beneath the vitelline membrane; and the haploid nucleus of the oocyte. Major steps in fertilization are illustrated in Figure 7-3.

The first step in fertilization involves penetration of the spermatozoon through the cumulus and corona radiata cells with its head sticking to the zona pellucida. Two enzymes, hyaluronidase and corona penetrating enzyme, aid this passage. Both are associated with the head of the sperm. Release of these enzymes is made possible by capacitation and the *acrosome reaction*. During the acrosome reaction, the outer acrosomal membrane and the plasma membrane of the head of the spermatozoon fuse. This fusion produces vesicles which release acrosomal enzymes (hyaluronidase and others) necessary for fertilization. The inner acrosomal membrane remains intact around the head of the spermatozoon. The acrosomal reaction can only occur in capacitated spermatozoa and is considered by some to be a part of capacitation.

During the second step the spermatozoon penetrates the zona pellucida

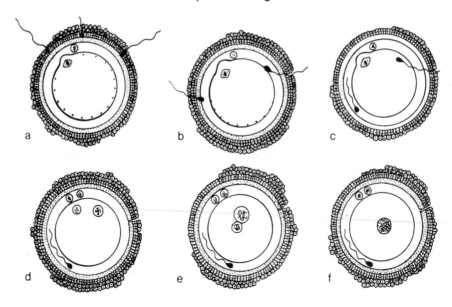

FIGURE 7-3.   Sequential steps in fertilization (Adapted from Austin and Bishop. 1957. *Biol. Rev.*, 32:296):

    (a)  Spermatozoon penetrates the cumulus and corona radiata cells sticking to the zona pellucida.

    (b)  A spermatozoon penetrates the zona pellucida and fuses with the vitelline membrane. The zona reaction is initiated as the cortical granules disappear.

    (c)  The spermatozoon is engulfed by cytoplasm in the oocyte. The vitelline block is evoked.

    (d)  The cytoplasm shrinks; a second polar body is pushed into the perivitelline space; the male and female pronuclei form.

    (e)  Syngamy occurs.

    (f)  The zygote is formed, completing fertilization.

and the plasma membrane of the sperm head fuses to the vitelline membrane. Acrosin, a trypsin-like enzyme associated with the head of the sperm, aids this penetration and a slit is left in the zona pellucida at the point of entry. Following passage of the spermatozoon, the *zona reaction* occurs. When the plasma membrane of the head of a spermatozoon fuses with the vitelline membrane, the cortical granules also fuse with the vitelline membrane and empty their contents into the perivitelline space. The spilling of the contents of the cortical granules into this space seems responsible for the zona reaction. The zona reaction guards against penetration of the zona pellucida by other spermatozoa. It is not an absolute safeguard, since other spermatozoa can sometimes be seen in the space between the zona pellucida and vitelline membrane. However, this does not happen frequently.

The spermatozoon then penetrates the vitelline membrane by phagocytosis and enters the cytoplasm. After this entry, the *vitelline block*, which is a second reaction to guard against fertilization by other spermatozoa, occurs. Cortical granules may participate in the vitelline block also. Upon entering the cytoplasm, the tail of the spermatozoon separates from the head. Mitochondria associated with the tail degenerate and other parts appear to dissolve in the cytoplasm. The cytoplasm shrinks and the second polar body is extruded. Both the male and the female pronuclei form. This involves an unfolding of the chromosomes in preparation for pairing. *Syngamy*, a merging of the pronuclei, then occurs. When syngamy is complete the zygote has been formed, thus completing fertilization.

## 7-4 Polyspermy

*Polyspermy* is a term used to describe fertilization by more than one spermatozoon. The result is a zygote with a triploid (3n) nucleus. Should an offspring result from such a fertilization all cells in the body would be triploid. Embryos with cells containing triploid nuclei will develop normally for a period of time then die and degenerate. Therefore, a major disadvantage of polyspermy is that it results in embryo loss. With the zona reaction and vitelline block guarding against polyspermy, it seldom occurs (1 to 2% in mammals). It is more likely to occur if the secondary oocyte is either aged or heated. Therefore, polyspermy is more likely if the animal is bred too late or if her body temperature is elevated by fever or high ambient temperature. Polyspermy could account for some of the reduction in conception rate and increase in early embryonic mortality under such conditions.

## 7-5 Aging of gametes

To achieve optimum conception during controlled natural breeding or AI it is essential that fertilization take place before either gamete has aged (see section 6-3). For species other than the horse, some loss in fertility of spermatozoa can be expected after about 24 hours following mating (Table 7-2). The fertile life of the ovum is usually less than 12 hours except possibly for the ewe. In Tables 7-3 and 7-4, the two-fold problem of aged gametes is illustrated. Fertilization rates are lower when the gametes have aged. Subsequently, both embryo and fetal death losses are higher whether the aged gamete is of male or female origin. It should be noted that the semen used in the research reported in Table 7-4 was diluted in egg yolk-citrate diluter and presumably stored at 5°C. Therefore, aging was slower than would occur in the reproductive tract of the cow. When one considers man's inability to estimate the time of ovulation due to lack of knowledge of the exact hour that estrus begins and the variability of the female, the critical nature of diligent management during the breeding season becomes apparent. Deviating from the optimum time to breed by only a few hours can result in a drop in conception rate.

TABLE 7-2.  Estimated fertile life of sperm and ova in farm animals

| Species | Fertile life in hrs | |
| | Spermatozoa | Ova |
|---|---|---|
| Cattle | 24-48 | 8-12 |
| Swine | 24-48 | 8-10 |
| Sheep | 30-48 | 16-24 |
| Horse | 72-120 | 6-8 |

Adapted from McLauren. *Reproduction in Farm Animals*, (3rd ed.). ed. Hafez, Lea and Febiger, 1974.

TABLE 7-3.  Effect of age of the ovum on fertility in cattle

| Hours from ovulation to insemination | Fertility observed at 2-4 days | | Fertility observed at 21-35 days | |
| | Total animals | Animals with fertile ova | Total animals | Animals with normal embryos |
|---|---|---|---|---|
| 2-4 | 4 | 75% | 4 | 75% |
| 6-8 | 4 | 75 | 10 | 30 |
| 9-12 | 5 | 60 | 13 | 31 |
| 14-16 | 4 | 25 | 8 | 0 |
| 18-20 | 5 | 40 | 6 | 17 |
| 22-28 | 1 | 0 | 11 | 0 |

From Nalbandov. *Reproductive Physiology of Mammals and Birds*. (3rd ed.) W. H. Freeman and Co. copyright © 1976.

TABLE 7-4.  The effect of the length of time of storage of extended semen on its fertility level and the difference between 1-mo. and 5-mo. nonreturns

| | Age of extended semen when inseminated in relation to day of collection | | | | |
| | Same day | 2nd day | 3rd day | 4th day | 5th day[a] |
|---|---|---|---|---|---|
| No. of inseminations | 12 | 726 | 756 | 970 | 56 |
| 1-mo. non-returns (%) | 58.3 | 67.0 | 62.8 | 54.3 | 57.2 |
| 5-mo. non-returns (%) | 50.0 | 57.0 | 50.7 | 41.5 | 39.3 |
| Difference (%) | 8.3 | 10.0 | 12.1 | 12.8 | 17.9 |

a 5th day or more.
From Salisbury, Bratton and Foote. J. Dairy Sci., 35:256. 1952.

**Suggested reading**

Adams, C. R. 1972. Aging and reproduction. *Reproduction in Mammals. 4. Reproductive Cycles.* ed. C. R. Austin and R. V. Short. Cambridge University Press.

Austin, C. R. 1972. Fertilization. *Reproduction in Mammals. 1. Germ Cells and Fertilization.* ed. C. R. Austin and R. V. Short. Cambridge University Press.

Baker, T. G. 1972. Oogenesis and ovulation. *Reproduction in Mammals. 1. Germ Cells and Fertilization.* ed. C. R. Austin and R. V. Short. Cambridge University Press.

Baker, T. G. 1972. Primordal germ cells. *Reproduction in Mammals. 1. Germ Cells and Fertilization.* ed. C. R. Austin and R. V. Short. Cambridge University Press.

Hafs, H. D. 1978. Ovigenesis, ovulation and fertilization. *Physiology of Reproduction and Artificial Insemination of Cattle.* (2nd ed.) ed. G. W. Salisbury, N. L. VanDemark and J. R. Lodge. W. H. Freeman and Co.

Mauleon, P. and J. C. Mariana. 1977. Oogenesis and folliculogenesis. *Reproduction in Domestic Animals.* (3rd ed.) ed. H. H. Cole and P. T. Cupps. Academic Press.

# Chapter

# 8

# Gestation

*Gestation* is the period of pregnancy. It starts with fertilization, which has been described, and ends with *parturition* (the birth process). The average length of the gestation period is 114 days for the sow, 148 days for the ewe, 281 days for the cow, and 337 days for the mare. Both individual and breed differences exist (Table 8-1). Gestations are a little longer when cows are carrying a male than when carrying a female. With twins, gestations are shorter in cows.

During the early part of gestation, the embryo floats free, first in the oviduct and then the uterus. Its nutrients are those which are stored in its own cytoplasm and those that can be absorbed from uterine milk. Only after placentation, the process by which the embryo becomes attached to the uterus, can the embryo derive nutrients and transfer waste products through maternal blood. The time of placentation after fertilization has been reported to start at 12 to 20 days in sows, 18 to 20 days in ewes, 30 to 35 days in cows, 50 to 60 days in mares. This time is more difficult to ascertain in the sow and mare with a diffuse placental attachment than in the ewe and cow with cotyledonary attachments. Early during the course of placentation, the placental attachments are quite fragile. In the cow between 30 to 35 days after fertilization, there will be 3 or 4 fragile cotyledonary attachments in the pregnant horn. Before 40 days fragile attachments will be present in both horns. There will be 40 to 50 cotyledonary attachments in both horns by 70 days and this increases to approximately 100 by the middle of the pregnancy. Before the embryo can derive benefit from placentation, organ development will have to occur to the point that the embryo's circulatory system is functional.

Spacing of embryos, which occurs before placentation in sows, is completed by 12 days. During this 12 day period embryos migrate freely from one side to the other. Even though about 55% of the ova come from the left ovary, embryos will be distributed equally between the two uterine horns of the sow after spacing is completed. Embryos migrate freely in mares, also, with the fetus frequently found on the side opposite to the corpus luteum of pregnancy. In cows and ewes, transuterine migration of embryos is less frequent than in mares.

There are three distinctive periods in the development of the conceptus. These are: (1) cleavage, (2) differentiation and (3) growth.

TABLE 8-1. Species and breed differences in gestation length

| Breed | Average length (Days) |
|---|---|
| Cow | |
| Ayrshire | 278 |
| Guernsey | 283 |
| Jersey | 279 |
| Holstein | 279 |
| Brown Swiss | 290 |
| Angus | 279 |
| Hereford | 284 |
| Shorthorn | 283 |
| Ewe | |
| Hampshire | 145 |
| Southdown | 145 |
| Merino | 151 |
| Mare | |
| Belgium | 335 |
| Morgan | 342 |
| Arabian | 337 |

**8-1**
**Cleavage**

After fertilization the zygote will divide and redivide many times without any increase in cytoplasm (Figure 8-1). The overall size may increase due to absorption of water, but the total cellular material will decrease. This process of cell division without growth is *cleavage*. The first cleavage will result in a 2-cell embryo. This is followed by additional cleavages resulting in 4-cell, 8-cell, 16-cell, 32-cell embryos, etc. When the embryo passes from the oviduct into the uterus, a ball of 16 to 32 cells will be contained within the zona pellucida. This structure, with cells too numerous to accurately count, is called a *morula*. During the next few days, fluid collecting in the intercellular spaces will push to the center forming the *blastocyst*, a structure with a fluid filled cavity (the blastocoele) surrounded by a layer of cells. The *inner cell mass*, a mound of cells on one side of the blastocyst which will eventually form the body of the embryo, can be identified. Cells in the blastocyst are not yet differentiated. Therefore, it is not possible to identify

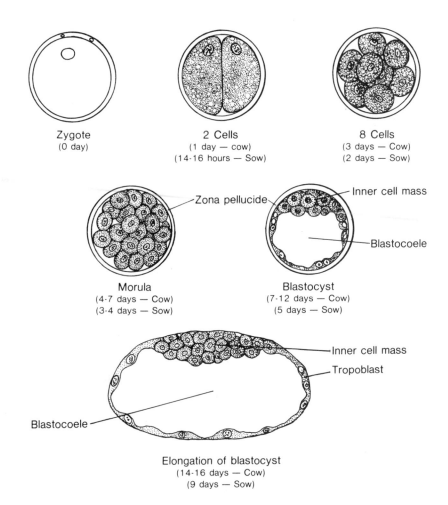

FIGURE 8-1. Specific cleavage stages at given times after fertilization in the cow (281 day gestation) and the sow (114 day gestation).

cells that will form a particular organ of the body. Near the end of the period of cleavage, the zona pellucida will weaken and disintegrate, permitting the blastocyst to elongate. After the period of cleavage, cell divisions will continue, but it will be cell divisions with growth.

The pattern of development during cleavage is similar for all farm species studied (see Table 8-2). The period of cleavage extends from fertilization to about 12 days for the cow, 10 days for the ewe and 6 days for the sow. A trend for more rapid development can be seen for species with shorter gestation periods. This trend will continue as the conceptus develops.

TABLE 8-2. Time comparisons from ovulation during cleavage for different farm species

| Species | 1 cell (hours) | 8 cell (days) | Blastocyst (days) | Enter uterus (days) |
|---|---|---|---|---|
| Cattle | 24 | 3 | 8 | 3.5 |
| Horse | 24 | 3 | 6 | 5 |
| Sheep | 24 | 2.5 | 7 | 3 |
| Swine | 14-16 | 2 | 6 | 2 |

Adapted from McLauren. *Reproduction in Farm Animals.* (3rd ed.) ed. Hafez, Lea and Febiger, 1974.

**8-2**
**Differentiation**

*Differentiation* might be called the true period of the embryo. It is a period when the cells are in the process of forming specific organs in the body of the embryo. Notable events during differentiation include the formation of the *germ layers, extraembryonic membrane,* and *organs.* In addition, rapid changes in relative size occur during differentiation.

**8-2.1**
**Germ layers**

The first evidence that differentiation has begun is the appearance of the germ layers (Figure 8-2). The *endoderm,* the innermost germ layer, first appears when a single layer of cells pushes out from the inner cell mass and grows around the blastocoele. The endoderm is the origin of the digestive system, liver, lungs, and most other internal organs (Table 8-3). The *mesoderm,* the middle germ layer, arises from the inner cell mass, pushing between the endoderm and *ectoderm.* The mesoderm is the origin of the skeletal system, muscles, circulatory system, and reproductive system. The

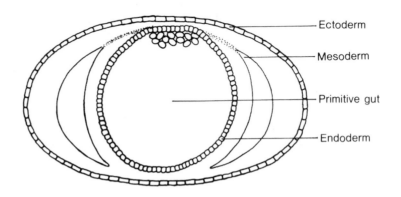

FIGURE 8-2. Germ layers as they appear in a section of an embryo two to three days after the start of differentiation.

ectoderm, the outer germ layer, is the origin of the nervous system, sense organs, hair, skin, mammary glands, and hooves.

TABLE 8-3. Certain organs that have been identified as forming from specific germ layers

| Germ Layer | Organs |
|---|---|
| Ectoderm | 1. Central nervous system |
| | 2. Sense organs |
| | 3. Mammary glands |
| | 4. Sweat glands |
| | 5. Skin |
| | 6. Hair |
| | 7. Hooves |
| Mesoderm | 1. Circulatory system |
| | 2. Skeletal system |
| | 3. Muscle |
| | 4. Reproductive systems (male and female) |
| | 5. Kidneys |
| | 6. Urinary ducts |
| Endoderm | 1. Digestive system |
| | 2. Liver |
| | 3. Lungs |
| | 4. Pancreas |
| | 5. Thyroid gland |
| | 6. Most other glands |

**8-2.2
Extraembryonic
membranes**

Soon after the appearance of the germ layers, formation of the extraembryonic membranes will begin (Figure 8-3). Two extraembryonic membranes, the *amnion* and *allanto-chorion*, will form during this period and function throughout the remainder of gestation. A third extraembryonic membrane, the *yolk sac*, is seen early during differentiation but will have disappeared by the end of this stage of development. The yolk sac contains an early source of nutrients for the developing embryo. As the yolk is depleted, the yolk sac regresses. A portion of the yolk sac is folded into the embryo, forming its primitive gut.

The amnion, the inner extraembryonic membrane, forms as the *trophoderm* (outer layer formed by fusion of ectoderm and mesoderm) folds around the embryo (see Figure 8-3), leaving an ectodermal layer on the inside of the amnion. The amnion contains fluids which suspend the embryo, protecting it and permitting its free growth. During the period of differentia-

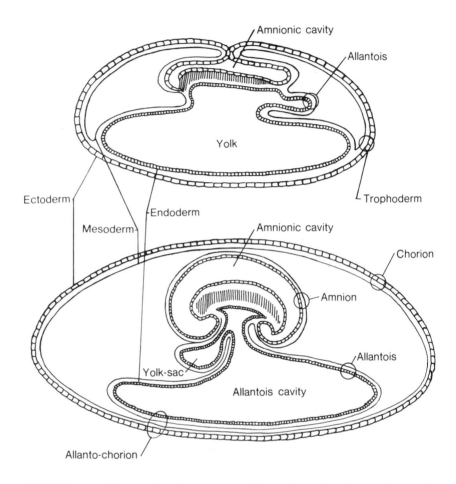

FIGURE 8-3. Progressive development of the extraembryonic membranes including fusion of the chorioamnionic folds and the allantois with the chorion. (Redrawn from Patten. 1964. *Foundations of Embryology*. McGraw-Hill.)

tion, the fluid in the amnion will make it turgid. The amnion can be palpated by way of the rectum in cows between 30 and 45 days, but its turgidity will not permit palpation of the embryo. Since the embryo is quite fragile during this early period, the turgidity helps maintain its shape and prevents injury during rectal palpation for pregnancy. As the amnion enlarges it becomes less turgid. By 60 days post-fertilization it will have softened enough for the fetus to be palpated by rectal palpation. The amnionic fluid will continue to bathe and suspend the fetus throughout gestation.

With formation of the amnion, the outer layer of extraembryonic membranes is called the chorion rather than the trophoderm. The allanto-chorion, the outer extraembryonic membrane, forms by a fusion of the chorion with the *allantois*. The beginning development of the allanto-chorion is illustrated in Figure 8-3. The allantois is a vascular membrane that is first seen as an outpouching of the hindgut. It connects to the embryonic bladder and contains fluids high in waste products. As the allantois enlarges, it fuses with the chorion until the allanto-chorion has completely formed around the amnion. The allanto-chorion becomes attached to the endometrium during placentation forming the *placenta*. After placentation, oxygen and nutrients from maternal blood pass through the placental attachments into the embryonic circulation which transports them to the developing embryo. Waste products, including ammonia and carbon dioxide from the embryo, are transported from embryonic blood through the placental attachments to maternal blood for elimination through the maternal system. Should the allanto-chorion not develop properly, the embryo would soon die from deprivation of oxygen and nutrients and/or buildup of toxic waste products.

**8-2.3**
**Organ formation**

As formation of the extraembryonic membranes progresses, cells within the inner cell mass are differentiating. A neural plate forms from ectodermal cells as the beginning of the central nervous system. The primitive brain and spinal cord are quickly discernible. The circulatory system develops rapidly from mesodermal cells and by day 16 in the sow and by day 22 in the cow an embryonic heartbeat can be detected. The liver, pancreas, lungs, and digestive system can be identified as they differentiate from endodermal cells. Within a few days, limb buds, which will form legs, a tail bud, and the lens of the eye can be identified. At this stage of development, embryos from most species (including human embryos) are so similar in appearance that they can not be distinguished as to species (Figure 8-4).

During the period when other organs are developing, the reproductive system will form. A dual duct system, the *Mullerian ducts* and the *Wolffian ducts*, will appear in all embryos. Sexual differentiation will not have occurred. If the embryo is a genetic female the pair of Mullerian ducts will develop into the female duct system. This includes the oviducts, uterus, cervix, and vagina. The Wolffian ducts will regress and disappear. If the embryo is a genetic male the pair of Wolffian ducts will develop into the male duct system. The male duct system includes the epididymis and vas deferens (see Chapter 3). The Mullerian ducts regress and disappear in the male.

The embryonic gonads arise on either side of the dorsal wall of the abdomen. These first appear as *genital ridges*, slight thickenings near the kidneys. The indifferent gonad soon differentiates into an inner medulla and outer cortex. In the genetic male, *primary sex cords* arise and extend into the medulla, which develops into the testes as the cortex regresses. Primordial germ cells migrate from the hindgut of the embryo into the primary sex cords, which will later differentiate into seminiferous tubules and the rete

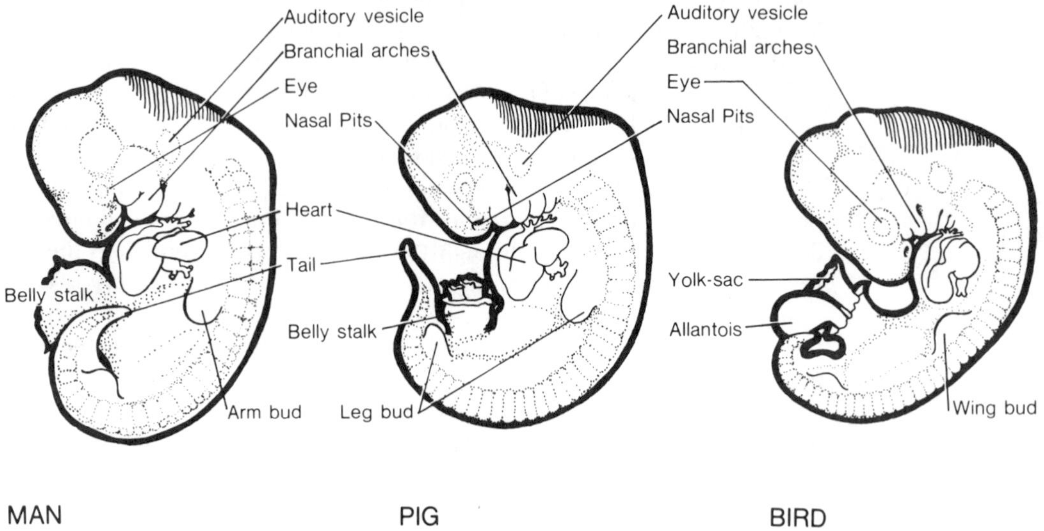

FIGURE 8-4. Embryos of man, pig, and bird at corresponding stages of development. (Redrawn from Patten. 1964. *Foundations of Embryology*. McGraw-Hill.)

testis. In the genetic female, appearance of primary sex cords will be followed by *secondary sex cords,* which arise from the surface epithelium and remain in the cortex, which develops into the ovaries. Primordial germ cells from the hindgut are incorporated into the secondary sex cords. These cords later break up into isolated clusters of cells called primary follicles. As described in Chapter 2, they consist of an oogonium surrounded by a single layer of follicular cells. A period of mitosis follows in the fetal period during which thousands of primary follicles are formed. The ovaries form later and develop more slowly than the testes. In the female, the primary sex cords and medulla regress. Appearance of the secondary sex cords is an early distinguishing feature of the female.

Both relative rate of growth and formation of organs proceed rapidly during differentiation (Table 8-4). The period of differentiation starts about day 7 in the sow, 11 in the ewe, and 13 in the cow. It will be completed by day 28 in the sow and by day 45 in the ewe and the cow. Primitive organs that appear early in the period of differentiation will be completely formed at the end of the period. The embryos at the end of this period will appear as miniature pigs, cows, sheep, or horses (see Figure 8-5). The head will have filled out and facial features will be distinct. Legs, hooves, and tail will have formed. In the male, a scrotum can be identified. In the female, mammary buds and the vulva can be seen. The function of organs is limited at the end of differentiation, but everything necessary for the development into functional organs will be present. The period of differentiation is a critical period. Anything that interferes with normal differentiation will not be corrected later in the gestation after differentiation is completed. Drugs that interfere with normal differentiation of legs or eyes will appear as defects in the offspring after parturition. Likewise, therapy with sex hormones may interfere with normal sexual differentiation.

TABLE 8-4. Developmental features in the cow and pig during differentiation

| Identifiable characteristics | First appearance | |
| | Cow (days) | Sows (days) |
| --- | --- | --- |
| Germ layer | 14 | 8 |
| Open neural tube | 20 | 13 |
| Fusion of chorio-amnionic folds | 18 | 16 |
| Heartbeat | 22 | 16 |
| Allantois prominent | 23 | 17 |
| Fore limb bud | 25 | 18 |
| Hind limb bud | 28 | 19 |
| Lens of eye | 30 | 21 |
| Placentation | 33 | 12 |
| Facial features distinct | 45 | 28 |

Adapted from Hafez, *Reproduction in Farm Animals*, (3rd ed.) Lea and Febiger, 1974.

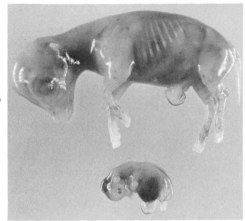

FIGURE 8-5. Fetal calf at 75 days of the gestation compared to a 42 day embryo.

**8-3**
**Fetal growth**

After differentiation is completed, the product of conception is called a *fetus* rather than an embryo. This portion of gestation, between the completion of differentiation and parturition, has been termed "the period of the fetus." The principal developmental feature of this period is growth.

Several landmarks in the development of the fetal calf have been identified. Calcification of bone matrix will start at about 70 days, with extensive bone formation having occurred by 180 days. Tooth formation will begin at about 110 days. Hair can be seen around the eyes and muzzle at 150 days, with hair covering the entire body by 230 days. In males, the testes will descend from the body cavity, through the inguinal canals into the scrotum. Descent of the testes will be completed by midgestation in bulls, but not until near the time of parturition in stallions.

The pattern of growth in the fetus is interesting. Relative growth rate is more rapid during early gestation than late gestation (see Table 8-5). Between 45 and 75 days, in the fetal calf, there will be over 1000% increase in size (e.g., 6 to 72 g). During the last 2 months of gestation a relative change of a little over 100% is seen (e.g., 18 to 40 kg). Even though the relative growth rate is slower during late gestation, over half of the total weight of the fetus at parturition is gained during the last 2 months. It is only during the last 2 months of the gestation that the mother must be given an added increment of nutrients to account for the growing fetus.

The weights of the fetal fluids, fetal membranes and the maternal uterus increase as gestation progresses. Just before parturition in the cow, the fetal fluids weigh about 15.5 kg and the fetal membranes weigh about 3.8 kg. The uterus will increase from about 1.0 kg to 10 kg during the course of pregnancy. Even with the ten-fold increase in the size of the uterus, the fetus and its associated fluids and membranes will account for about 85% of the total weight of the uterus and its contents.

**8-4**
**Twinning**

Twinning sometimes occurs in monotoccus species. It occurs more frequently in sheep than in cattle. The percent lambs per lambing may exceed 150% for some flocks and lambing rates of 120 to 140 are not uncommon. Twinning is not considered undesirable in sheep, in that it increases the number of lambs that are weaned in a given year. Double ovulations occur in about 25% of all estrous cycles in the mare and is a problem because abortion usually occurs with twins. Many horse breeders palpate mares to avoid breeding when two large Graafian follicles are present during a period of estrus. If rupture of one follicle precedes rupture of the other by as much as 2 days, mares can be bred for fertilization of the oocyte in the second follicle.

Twinning rates in cattle are relatively low, ranging from 0.5 to 4% for different breeds. In some herds, rates of 8 to 10% have been reported. Twinning rates for Brown Swiss and Holsteins are higher than for Jerseys and Guernseys, and dairy cows have higher twinning rates than beef cows. The twinning rate in most beef breeds is less than 1%. Twinning has not

TABLE 8-5. Weight changes of the bovine uterus and its contents during pregnancy

| Stage of gestation | Total uterus and contents | Embryo or fetus | Amnionic fluids | Fetal membranes | Empty uterus |
|---|---|---|---|---|---|
| days | kg | gm | gm | gm | kg |
| 0-30 | .9 | .5 | — | 4.5 | .9 |
| 31-60 | 1.6 | 5.9 | 181.6 | 49.5 | 1.4 |
| 61-90 | 2.3 | 72.6 | 590.2 | 149.8 | 1.5 |
| 91-120 | 4.0 | 531.4 | 1600.0 | 258.8 | 1.7 |
|  |  | kg | kg | kg |  |
| 121-150 | 10.1 | 1.6 | 5.0 | .7 | 2.8 |
| 151-180 | 14.6 | 3.8 | 5.5 | 1.3 | 4.1 |
| 181-210 | 23.8 | 9.5 | 6.4 | 2.5 | 5.5 |
| 211-240 | 37.4 | 17.7 | 10.0 | 2.4 | 7.3 |
| 241-270 | 53.8 | 28.6 | 11.8 | 3.4 | 10.0 |
| 271-300 | 67.8 | 39.9 | 15.4 | 3.8 | 8.6 |

From *Physiology of Reproduction and Artificial Insemination of Cattle.* G. W. Salisbury, N. L. Van Demark, and J. R. Lodge, W. H. Freeman Co., copyright © 1978.

been considered desirable in cattle because of increased incidence of retained placentae, reduction in future reproductive efficiency, weaker calves that are more difficult to raise, and reduced milk production by cows after twinning.

The heritability of twinning is low. A higher incidence of twinning has been reported for certain cow families, but long term selection studies to increase twinning have not greatly increased the twinning rate. Twinning seldom occurs in *primiparous* (first gestation) females. The incidence of twinning increases with age for the next several years, then drops off again. These age effects are reported for both cattle and sheep. Increasing the level of nutrition will increase the incidence of twinning in sheep. While not clearly demonstrated, level of nutrition may be a factor in cattle and could account for some of the difference between dairy and beef breeds. Seasonal effects on twinning have been reported, but may be related to seasonal changes in available feed. Increased twinning can be expected after hormone therapy for cystic ovaries or for other reproductive disorders.

Most twins are of the *dizygous type*. That is, they result from ovulation of two oocytes during the same estrous cycle. These oocytes are fertilized and eventually implanted in the uterus where they are carried until parturition. They may be the same sex or opposite in sex. They are no more alike than siblings with the same parents born from different gestations.

Some twins are *monozygous*, resulting from fertilization of a single oocyte. Monozygous twins are always the same sex and are genetically and phenotypically identical, except that one is frequently larger than the other.

The means by which a single zygote can result in twins is not known. Theories that have been advanced include both separation of the cells after the first cleavage with each cell developing independently and formation of two inner cell masses within the same blastocyst (Figure 8-6). Other theories involving aberrations in differentiation have been advanced. About 8 to 10% of all twin births in cattle are monozygous. About 30% of all human twins are monozygous with a higher loss during the gestation than with dizygous twins. Embryo loss and abortion may account for the lower percentage of monozygous twins in cattle.

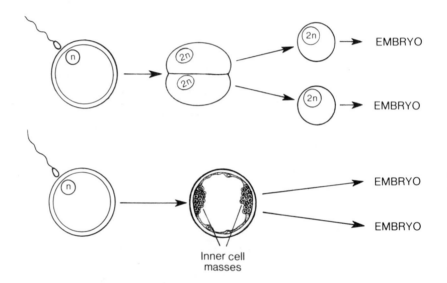

FIGURE 8-6. Two theories on development of monozygous twins.

## 8-5 Hormones important to gestation

Maintenance of pregnancy is largely dependent on a proper balance of hormones. This is evidenced by the fact that disturbance of the normal balance frequently results in abortion. Maintenance of the proper balance of hormones throughout the gestation period is dependent on an interplay between the mother, the placenta and possibly, the fetus.

Progesterone is the hormone identified to have the most dominant role in maintenance of pregnancy. High levels of progesterone will decrease the tone of the myometrium and inhibit uterine contractions. It does this by blocking the action of other hormones or drugs on the myometrium. Also, high progesterone concentrations stop cyclic estrus by preventing the release of gonadotrophins. Progesterone is produced by both the corpus luteum and the placenta. Cows and sows are dependent on the corpus luteum as a source of progesterone for most of the gestation period. Ewes are dependent on the placenta during the last half of pregnancy. This reproductive function is more complex in mares (Figure 8-7). The corpus luteum which forms at the ovulation site is active for 150 to 180 days. Accessory corpora lutea form on the ovaries when pregnant mare serum gonadotrophin (PMSG) stimulates the formation and luteinization of small tertiary follicles. These form between 35 and 40 days post-fertilization and actively secrete progesterone from 150 to 180 days. After 150 to 180 days post-fertilization, progesterone necessary for maintenance of pregnancy in the mare must come from the placenta. All corpora lutea regress at this time. A drop in progesterone at this time corresponds to an increase in estrogen. If the synthesis of progesterone stops in any species the pregnancy usually terminates within a few days.

Relaxin, a polypeptide produced by the corpus luteum and the placenta, appears important during gestation. Concentrations of relaxin increase during gestation. The levels are much higher during late gestation than during early gestation. Primarily, relaxin appears to soften connective tissue, which permits uterine muscles to stretch to accommodate a growing fetus.

Concentrations of estrogens are low during early gestation but increase during middle and late gestation. In mares estrogen levels are quite high

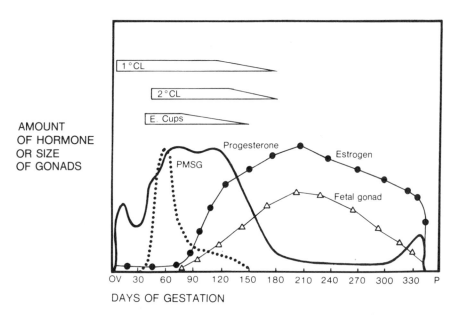

FIGURE 8-7. Progesterone, estrogen, and PMSG concentrations during pregnancy in the mare. (Stabenfelt and Hughes. 1977. *Reproduction in Domestic Animals.* (3rd ed.) ed. Cole and Cupps. Academic Press.)

during the last half of gestation. The primary source of these estrogens is the placenta. The principal function of the estrogens during gestation appears to be their synergistic action with progesterone in developing and preparing the mammary glands for synthesis of milk after parturition.

Other primary hormones of reproduction do not appear to have a dominant role during gestation. Low concentrations with little variation are usually found in the blood. Normal function of the thyroid, parathyroid, adrenal cortex and other endocrine glands which produce secondary hormones of reproduction are important to maintenance of a metabolic state in the mother which permits proper embryonic and fetal development.

A number of endocrine changes occur near the end of the gestation period. These will be discussed in the next chapter as they relate to parturition.

**Suggested reading**

Arey, L. B. 1974. *Developmental Anatomy.* (7th ed.) W. B. Saunders Co.

Eckstein, P. and W. A. Kelly. 1977. Implantation and development of the conceptus. *Reproduction in Domestic Animals.* (3rd ed.) ed. H. H. Cole and P. T. Cupps. Academic Press.

Gomes, W. R. 1978. Gestation. *Physiology of Reproduction and Artificial Insemination of Cattle.* (2nd ed.) ed. G. W. Salisbury, N. L. Van-Demark and J. R. Lodge. W. H. Freeman and Co.

Liggins, G. C. 1972. The fetus and birth. *Reproduction in Mammals. 2. Embryonic and Fetal Development.* ed. C. R. Austin and R. V. Short. Cambridge University Press.

McLauren, A. 1972. The embryo. *Reproduction in Mammals. 2. Embryonic and Fetal Development.* ed. C. R. Austin and R. V. Short. Cambridge University Press.

Moore, K. L. 1977. *The Developing Human.* (2nd ed.) W. B. Saunders Co.

Patten, B. M. 1964. *Foundations of Embryology.* (2nd ed.) McGraw-Hill.

Short, R. V. 1972. Sex determination and differentiation. *Reproduction in Mammals. 2. Embryonic and Fetal Development.* ed. C. R. Austin and R. V. Short. Cambridge University Press.

Swett, W. W., C. A. Mathews and M. H. Fohrman. 1948. Development of the fetus in the dairy cow. *U.S. Dept. Agr. Tech. Bull.* 964.

# Chapter

# 9

# Parturition and postpartum recovery

**9-1**
**Overview of the parturition process**

Parturition is the birth process. It begins with softening and initial dilation of the cervix along with the start of uterine contractions. It ends when the fetus and its associated placental membranes are expelled. The time required for parturition varies among individuals and among species (Table 9-1).

Parturition can be divided into three stages. The first stage of parturition ends with complete dilation of the cervix and entry of the fetus into the cervix. This stage usually takes from 2 to 6 hours in cows and ewes. More time is required in sows and less time in mares. The second stage ends with expulsion of the fetus. In all species less time is required for this stage than for the first stage, usually taking no more than 2 hours in cows and ewes. A similar time is required in sows, but the time varies with the size of the litter. The second stage is completed in 15 to 20 minutes in mares. In the mare, fetal expulsion must be rapid. The placental membranes become separated from the uterus and the foal will suffocate if a longer time is required.

TABLE 9-1. Average time required for the three stages of parturition for different species of farm animals

| Animal | Stage | | |
| --- | --- | --- | --- |
| | 1 | 2 | 3 |
| | Hours | Hours | Hours |
| Cow | 2-6 | 0.5-2 | 4-5 |
| Ewe | 2-6 | 0.5-2 | 0.5-8 |
| Sow | 2-12 | 1-4 | 1-4 |
| Mare | 1-4 | 0.15-0.5 | 0.5-3 |

Compiled from Roberts. *Veterinary Obstetrics and Genital Diseases*. Pub. by author. Ithaca, N.Y. 1971.

The first two stages take longer in first-gestation females of all species than in *multiparous* (second or later gestation) females.

Expulsion of the placenta occurs during the third stage. This may occur within 30 minutes of the expulsion of the fetus, but is more likely to occur 3 to 5 hours later. Our understanding of the parturition process as it is regulated by the endocrine system will be discussed.

**9-2
Approaching
parturition**

Signs of approaching parturition can be seen during the last month of the gestation. Careful evaluation of these signs will indicate when the female will need to be closely observed.

9-2.1
Rotation to birth
position

For monotoccus species such as the ewe, mare, and cow, the first signs of approaching parturition may start with the rotation of the fetus into the birth position. During most of the gestation in these species, the fetus will be on its back with its feet pointing up. After rotation into the birth position, it will be resting on its thorax or abdomen with its forefeet positioned at the uterine end of the cervix and its nose resting between the forefeet (Figure 9-1). Parturition proceeds more easily when the fetus is in this position,

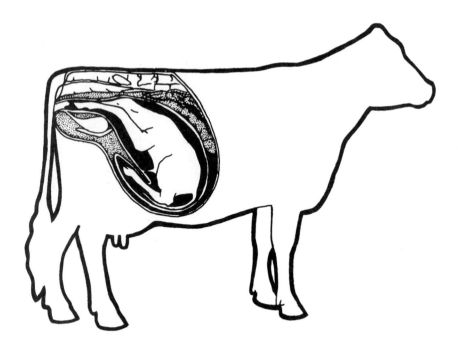

FIGURE 9-1.    Normal birth position in cows which is assumed near the end of the gestation period. (G. W. Salisbury, N. L. Van Demark, and J. R. Lodge. *Physiology of Reproduction and Artificial Insemination of Cattle.* (2nd ed.) W. H. Freeman and Co., copyright © 1978.

except in pigs where both front and back delivery proceeds with equal ease.

Abnormal positions are sometimes seen and occur more frequently with twins. The incidence of abnormal presentations is about 5% in cattle with a variety of abnormalities possible (Figure 9-2). This may range from one leg to both legs or the head turned back to several breach positions with the tail pointed toward the cervix. In those cases where the normal birth position is not assumed, assistance in delivery of the fetus is usually required. This involves repositioning the fetus to the normal position. The process of repositioning the fetus may break the umbilical cord. Should this happen, the fetus will suffocate if it is not quickly removed from the uterus. Assistance from a veterinarian is advised in cases of abnormal presentations. The cause (or causes) of rotation of the fetus into normal or abnormal positions is not known. The action of relaxin in expanding the pelvis may facilitate this movement.

**9-2.2
Mammary gland changes**

Growth of the mammary glands can be seen during the latter part of the gestation. This is likely caused by the synergistic actions of estrogens and progestins, which stimulate development of both ducts and secretory tissue in the mammary glands. As parturition approaches the mammary glands will enlarge as they fill with milk. Synthesis of milk is a function of prolactin in synergism with other hormones (see Chapter 10). As oxytocin is released during labor, milk letdown will occur frequently causing milk to leak from the teats.

**9-2.3
Other changes**

With parturition imminent, relaxin synergizing with estrogen will cause further expansion of the pelvis, enlarging the birth canal to facilitate passage of the fetus. A sinking of the ligaments around the tailhead will make the tailhead appear more prominent. The vulva will soften and become swollen. Mucus may be seen stringing from the vulva as estrogen causes the epithelial cells of the cervix to secrete new mucus, loosening the mucous plug. A "nesting instinct" thought to be stimulated by prolactin will be seen. The sow will actually build a nest. While cows do not build nests, they will try to leave the herd, seeking a place to remain hidden during parturition.

**9-3
Parturition**

Approximately 2 days before the start of parturition a rapid sequence of changes in hormone levels involving both the fetus and mother can be seen (Figure 9-3).

**9-3.1
Hormonal initiation**

The hormone pattern established during the latter part of the gestation period sets the stage for parturition. High levels of progesterone, relaxin, and estrogen are seen. On the basis of the present evidence it appears that the triggering mechanism initiating parturition is a release of cortisol by the fetus. In cows and sows, the rise in fetal cortisol causes production of estrogens by the placenta resulting in still higher levels which initiate release of

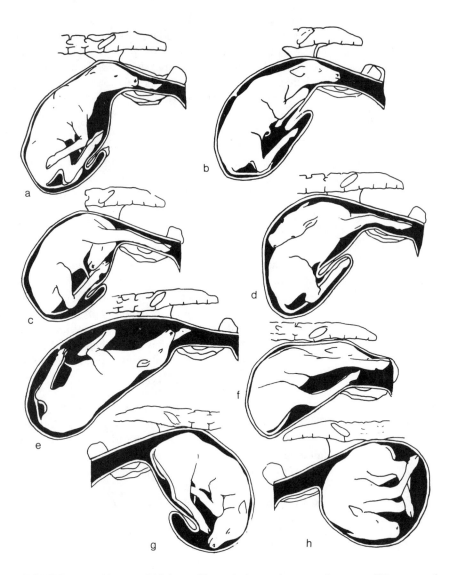

FIGURE 9-2. Abnormal birth positions that may be seen in cows. *(Diseases of Cattle. 1942. USDA Special Report.)*

PGF₂α from the uterus. PGF₂α causes regression of corpora lutea and a marked drop in progesterone concentration. In ewes, a similar sequence is seen, but the order differs. The placenta is the major source of progesterone in the ewe during late pregnancy. It appears that the rise in fetal cortisol causes changes in placental enzymes which result in conversion of placental progesterone to estrogens. Placental estrogens cause release of PGF₂α from

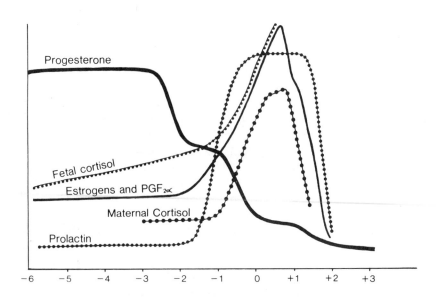

FIGURE 9-3. Relative changes in hormone concentration near the time of parturition in the ewe. Missing from this diagram but important to the regulation of parturition is oxytocin and relaxin. (Based on data in Chamley *et al.* 1973. *Biol. Reprod.*, 9:30 and Challis *et al.* 1977. *Biol. Reprod.*, 16:57.)

the uterus of the ewe, but the decrease in progesterone is seen before the rise in PGF2α. Thus, for all species studied fetal cortisol will initiate a sequence of changes that includes a sharply reduced level of progesterone with increased concentrations of estrogen and PGF2α in the blood serum of the mother prior to the start of parturition. Oxytocin is released as movement of the fetus stimulates sensory nerves in the cervix. An increase in maternal cortisol near the start of parturition is probably due to the stress of parturition on the mother and is not involved in the regulation of parturition. The surge of prolactin is related to milk synthesis and not to parturition.

**9-3.2
Regulation of
the
physiological
events**

The main physiological events in parturition are (a) dilation of the cervix which permits the fetus to pass and (b) uterine contractions which forcefully expel the fetus and placental membranes.

The initial dilation of the cervix is caused by relaxin as it synergizes with rising levels of estrogen. Working together, these hormones soften the cervix and cause the epithelial cells to secrete mucus. Further dilation occurs as uterine contractions force the chorion and later the amnion against the cervix. Complete dilation occurs as the fetus is forced into the cervix. The chorion may break during this process, but the amnion usually does not break until after the fetus enters the cervix.

A number of factors contribute to the initiation and continuation of uterine contractions which occur concurrently with dilation of the cervix and then continue for several hours after expulsion of the fetus. The reduced level of progesterone removes the inhibition to contractions from the myometrium. At the same time, rising levels of estrogens sensitize the myometrium, making it more responsive to stimulatory agents.

The initial uterine contractions are probably caused by $PGF_2\alpha$ as it is released from the endometrium by rising estrogen levels. These early contractions are weak and irregular, occurring at about 15-minute intervals. They become progessively stronger and more frequent as parturition proceeds. It is likely that $PGF_2\alpha$ is augmented by oxytocin which causes stronger contractions. The movement of the fetus into the cervix stimulates sensory nerves, which causes the release of oxytocin from the posterior pituitary. Fetal motility due to anoxia (low oxygen) may be another factor contributing to stronger contractions near the completion of the stage when the fetus is expelled. As uterine muscles contract reducing the blood flow to the fetus, its oxygen supply will be reduced causing the increased activity associated with anoxia. The mechanical movement of the fetus pushing against uterine muscles that are in a sensitized and contracting state evokes a stronger contractile response from these muscles. Just before expulsion of the fetus, uterine contraction will be regular, strong, and frequent, occurring at about 2-minute intervals and lasting for approximately 1 minute. Contraction of abdominal muscles will aid final expulsion. Following expulsion of the fetus the strong uterine contractions subside. However, diminished contractions will continue for another 1 to 2 days. The continued contractions are responsible for expulsion of the placental membranes as well as fluids and fragments of placental tissues that may remain in the uterus.

## 9-4
## Care of the newborn

Reproduction is not successfully completed unless the neonate (newborn) survives. The first few hours after expulsion from the uterus is critical to its survival. It has been transferred from a germ-free environment in which its temperature regulation, nutrient, and oxygen requirements have been dependent on the maternal system to an environment where it must survive independently.

Immediate changes must occur in the circulatory and respiratory systems. The umbilical artery has been taking blood saturated with carbon dioxide to the placenta where the carbon dioxide is exchanged for oxygen and then returned to the fetal heart through the umbilical vein to be pumped throughout the fetus. Fetal lungs have been present but have not been functioning as a respiratory organ. Most of the blood reaching the fetal heart bypasses the lungs by diversion into a parallel system that functions in the fetus. As the fetus passes through the vulva, its umbilical cord breaks, severing the neonate from the maternal system. Immediate survival depends on the cessation of circulation in the umbilical vessels along with the routing of blood through the lungs rather than through the parallel system that functioned in the fetus. In addition, respiration must start to oxygenate the blood as it circulates through the lungs. Increasing levels of carbon dioxide stimulate the respiratory center in the brain, thus initiating respiration.

The neonate is more susceptible to extremes in environmental temperature than an older animal whose thermoregulatory mechanisms have had time to adjust to fluctuating environmental temperatures. In cold environments, the rectal temperature of some species has been reported to drop several degrees within a few hours after birth. In both extremely hot and extremely cold environments neonatal survival is reduced.

A time for adjustment to absorption of nutrients from the digestive system along with their metabolism and utilization is needed also. A neonate has a relatively large amount of glycogen stored in its liver and muscles to be utilized during this adjustment period. Climate stress speeds the depletion of these reserves.

The neonate's immune system has not been challenged. Therefore, it does not have circulating antibodies to ward off diseases. Maternal antibodies do not pass through the placental barrier to fetal blood in cattle, sheep, swine, or horses. Maternal antibodies can be absorbed through the intestines if the newborn receives a meal of *colostrum* (the first milk) from its mother soon after parturition. There is evidence that the newborn loses the ability to absorb maternal antibodies through the intestines as early as 6 hours after birth. Stress from high temperature reduces absorption of maternal antibodies. Other stresses may also reduce antibody absorption.

An animal caretaker who is aware of the challenges to the survival of the neonate will likely take steps to increase the chances of survival. When the umbilical veins and arteries break, they will be retracted into the stump of the umbilical cord. Very little loss of blood will occur. The umbilical stump should be painted with an antiseptic solution such as iodine to lessen the chance of infection. The respiratory passages should be checked for pieces of placental membrane or mucus that could prevent respiration. If breathing has not started, it might be stimulated by slapping the neonate or by giving artificial respiration. One should not expect a high success rate in cases where respiration has not started spontaneously.

In cold weather, the newborn should be dried and if possible placed in a clean, dry shelter. In the summer, a shade will be beneficial to survival. If the newborn has not nursed within the first hour, it should be assisted in obtaining a meal of colostrum, to provide both nutrients and antibodies. If the mother dies during parturition, colostrum from another female of the same species should be provided in a pail or bottle with a nipple. Normal milk will provide most needed nutrients, but will not contain the needed antibodies (Chapter 10).

**9-5**
**Retained**
**placentae**

During normal parturition in the sow and mare the placental membranes become separated from the endometrium during fetal expulsion. The continued contractions of the uterus expel these membranes with retention seldom occurring. In ewes and cows, removal of blood from the cotyledons and continued contractions of the uterus loosen the chorionic villi from the caruncles with expulsion occurring shortly thereafter. Retention is rare in ewes but sometimes occurs in cows.

Retained placentae (also called retained afterbirths) will occur in from 5 to 15% of the parturitions in healthy herds of cows. They occur when separation of the chorionic villi from the caruncles is delayed. Many of these can be associated with short gestations (270 to 275 days) and twin births which result in shorter gestation periods. They are more likely to occur in high milk producers than low producers, which may account for the higher incidence in dairy cows than is seen in beef cows. Also, retained placentae are more common after difficult births (dystocia) than after uncomplicated births.

In herds where the incidence is over 15 to 20% a problem is indicated. Nutritional deficiencies of either selenium or vitamin A have caused a much higher rate of placental retention. If selenium injected a few weeks before parturition or vitamin A injected or supplemented through the diet should return the retention rate to the expected range, the cause of the problem will have been identified. If placental retention is associated with a higher than expected abortion rate, reproductive diseases may be involved (see Chapter 25). Assistance from a veterinarian will be needed in identifying and treating disease problems.

If the placenta of a cow has not been expelled 24 hours after expulsion of the fetus, it will likely be retained another 5 or 6 days. The decaying placental tissues make excellent media for microbial growth. Uterine infections occur in conjunction with most retained placentae. Some infections may not be eliminated for several weeks. Lower fertility after retention is a problem, with some cows never conceiving again. The stress associated with this problem will reduce milk production below anticipated levels.

Several alternatives are available in the management of the retained placenta problem. None are as satisfactory as measures that would prevent the problem. One alternative is to do nothing. Even though some cows recover without treatment and reproduce again, this is not a recommended approach. Problems from uterine infections and toxemia occur too frequently.

Another alternative is manual removal 48 to 72 hours after parturition, followed by intrauterine treatment with antibiotics. This method was used by veterinary practitioners for many years. Because of complications which reduced reproductive efficiency, it is no longer recommended. Manual separation of the chorionic villi from the caruncles frequently causes tears in the uterus that create reproductive problems.

Two approaches are now being recommended. One involves placing antibiotic boluses into the uterus daily until the placenta is expelled. The other involves a single large volume (4-liter) infusion of an antibiotic solution into the uterus 24 to 36 hours after parturition. Both methods give satisfactory results. The latter might be preferred because only one treatment is needed. Systemic (intravenous or intramuscular) treatment with antibiotics is not recommended unless the animal's temperature is elevated. Treatment with diethylstilbestrol (synthetic estrogen) and/or oxytocin may be of benefit if the placental membranes have been loosened from the endometrium but have not been expelled because uterine contractions have stopped. Indiscriminate use of hormones should be avoided. A higher incidence of cystic follicles has been reported after diethylstilbestrol treatment.

**9-6**
**Postpartum**
**recovery**

Following parturition, the next successful gestation depends on both return to normal estrus and return of the uterine environment to a state that will support another pregnancy. Sows frequently have a "farrowing estrus" a few days after parturition. If bred at this estrus, conception is very low, due to failure to ovulate. Mares have a "foaling estrus" 8 to 15 days after parturition. They should be bred at this time only if a careful examination indicates a complete recovery from gestation. A bruised or torn cervix, vaginal discharge, or lack of uterine tone are good reasons not to breed at foaling estrus. Normally, mares will return to estrus again by 30 days postpartum.

**9-6.1**
**Postpartum**
**ovulation and**
**estrus**

After parturition, many dairy cows will ovulate in 20 to 30 days. However, a high percentage of these ovulations will occur without evidence of estrus. These are frequently called "silent heats" or "quiet ovulations." When these cows ovulate again at 40 to 50 days, the majority will show signs of estrus. Such cows are less likely to become reproductive problems than those that have an extended period of *anestrus* (absence of cycling). Several factors can extend postpartum anestrus.

Nursing the young will delay return to estrus. Cows nursing a calf will likely go twice as long as cows not nursing before returning to estrus. The cause of the delay is not known, but the stress of *lactation* (milk production) does not appear to be a major factor. Dairy cows being milked twice daily are apparently not delayed. On the other hand, milking a dairy cow four times per day has delayed return to estrus. A similar effect is seen in sows. After the farrowing heat most sows stay in anestrus until their pigs have been weaned (5 to 8 weeks). If their pigs are removed after 1 week they will return to estrus in 3 weeks.

Low nutrition, either during the gestation or after parturition, will delay return to estrus. Quiet ovulations are also more frequent for cows on a low plane of nutrition. Cows most affected seem to be those in thin condition at time of calving. A combination of low nutrition and suckling will compound the problem. Many beef cows which calve in thin condition and nurse their calves will stay in anestrus well beyond 100 days. Some may not have a calf the next year because of failure to show estrus and ovulate while running with bulls during the next breeding season.

Other factors which may extend postpartum anestrus include infectious diseases, metabolic disorders, uterine infections, and other health problems.

## 9-6.2
## Involution of the uterus

Involution of the uterus implies the return to the normal nonpregnant state. This includes return to the nonpregnant size as well as recovery and repair of the endometrium. Return to nonpregnant size is quite rapid in sows and mares, being completed in less than 2 weeks. In ewes the uterus returns to the nonpregnant size in approximately 2 weeks, but another 2 weeks is needed for complete recovery of the endometrium.

Uterine involution has been studied more extensively in cows than in other species. This is partly because it is easy to follow the course of involution by rectal palpation. Criteria for involution include (a) return of the uterus to the pelvic area; (b) return to the nonpregnant size; and (c) recovery of normal uterine tone. Using these criteria, uterine involution in cows following uncomplicated parturitions requires about 45 days. Histological studies have shown that another 15 days may be required before the endometrium is histologically normal.

The authors have found that uteri of cows return to the nonpregnant size in about 30 days. The tone of the nonpregnant horn may be normal at that time. However, another 2 weeks may be needed for normal tone to return to the pregnant horn. Periods of estrus seem to speed the return of normal tone. It should be noted that one uterine horn may be larger than the other after complete involution. Also, in older cows, uteri may not return to the pelvic area. Therefore, tone as determined by rectal palpation is the more precise indicator of involution. Following placental retention and/or uterine infection, involution of the uterus may be delayed by several weeks.

## Suggested reading

Casida, L. E. 1971. The postpartum interval and its relation to fertility in cow, sow and ewe. *J. Anim. Sci.*, 32 (Supp. I):66.

Challis, J. R. G., J. K. Kendall, J. S. Robinson and G. D. Thornburn. 1977. The regulation of corticosteroids during late pregnancy and their role in parturition. *Biol. Reprod.*, 16:57.

Chamley, W. A., J. M. Buckmaster, M. E. Cerini, I. A. Cummings, J. R. Goding, J. M. Obst, A. Williams and C. Winfield. 1975. Changes in the level of progesterone, corticosteroids, estrone, estradiol-17β, luteinizing hormone and prolactin in the peripheral plasma of the ewe during late pregnancy and at parturition. *Biol. Reprod.*, 9:30.

First, N. L. 1979. Mechanisms controlling parturition in farm animals. *Beltsville Symposia in Agricultural Research. 3. Animal Reproduction.* ed. H. W. Hawk. Allanheld, Osmun and Co. Publishers; Halsted Press, a division of Wiley & Sons.

Gier, H. T. and G. B. Marion. 1968. Uterus of the cow after parturition: Involution changes. *Amer. J. Vet. Res.*, 29:38.

Inskeep, E. K. 1979. Factors affecting postpartum anestrus in beef cattle. *Beltsville Symposia in Agricultural Research. 3. Animal Reproduction.* ed. H. W. Hawk. Allanheld, Osmun and Co. Publishers; Halsted Press, a division of Wiley & Sons.

Korenman, S. G. and J. F. Krall. 1977. The role of cyclic AMP in the regulation of smooth muscle cell contractions in the uterus. *Biol. Reprod.*, 16:1.

Morrow, D. A., S. J. Roberts, K. McEntee and H. G. Gray. 1966. Postpartum ovarian activity and uterine involution in dairy cattle. *J. Amer. Vet. Med. Assoc.*, 149:1596.

Ryan, K. J. 1977. New concepts in hormonal control of parturition. *Biol. Reprod.*, 16:88.

Salisbury, G. W., N. L. VanDemark and J. R. Lodge. 1978. Parturition. *Physiology of Reproduction and Artificial Insemination of Cattle.* (2nd ed.) W. H. Freeman and Co.

Thornburn, G. D., J. R. G. Challis and W. B. Currie. 1977. Control of parturition in domestic animals. *Biol. Reprod.*, 16:18.

Wiltbank, J. N., W. W. Rowden, J. E. Ingalls, K. E. Gregory and R. M. Koch. 1962. Effect of energy level on reproductive phenomena of mature Hereford cows. *J. Anim. Sci.*, 21:219.

# Chapter

# 10

# Lactation

Lactation is the production of milk. Lactation has the primary purpose of providing nourishment for young offspring in most species. When these offspring no longer nurse their mothers, lactation stops. In the cow, sheep, goat, and a few other species, selection and breeding for higher milk production has provided an excess of milk over that needed by the young. This excess permits milk from these species to become an important part of the diet of both young and adult humans throughout the world.

A second purpose of lactation is to provide antibodies for the newborn. These are supplied through colostrum and can be absorbed by the neonate during the first few hours of life. These antibodies provide the newborn with its first internal resistance to diseases.

**10-1
Structure of
mammary
glands**

The *mammary gland* is the organ which produces milk. When comparing species, certain anatomical and morphological differences can be noted (Table 10-1).

TABLE 10-1.   Comparison of the mammary glands of various species

| Species | No. of glands | No. of teats | Streak canals in teat | Position of glands |
|---------|---------------|--------------|------------------------|---------------------|
| Cow | 4 | 4 | 1 | Inguinal |
| Mare | 2 gland complexes | 2 | 2 | Inguinal |
| Ewe | 2 | 2 | 1 | Inguinal |
| Doe | 2 | 2 | 1 | Inguinal |
| Sow | 4-9 pair | 4-9 pair | 2 | Abdominal |

**10-1.1
Anatomy**

The cow has four mammary glands (quarters) fused together into a single structure, the *udder* (Figure 10-1). They are located in the inguinal region with two glands lined up on either side of the midline. Each gland has a single teat. A *streak canal* through the teat permits removal of milk which has been produced and stored in that gland. Even though fused together, each gland is a separate unit. For example, dye injected into a single teat will be found only in the gland drained by that teat.

The mammary system of the mare appears as two glands with two teats. However, as in the cow, the mare has four separate areas of secretory tissue located in the inguinal region on either side of the midline. On either side of the midline, two secretory areas will be fused into a single gland complex which is drained by a single teat (Figure 10-2). Each teat has two streak canals which drain the separate secretory areas.

Both the ewe and doe (goat) have two mammary glands with one teat for each gland. These glands are fused together with one on either side of the midline in the inguinal region. Each teat has a single streak canal.

The sow has from four to nine pairs of mammary glands, which are located on either side of the midline along the entire abdominal wall. Each gland has a teat with two streak canals which drain separate secretory areas in the individual glands.

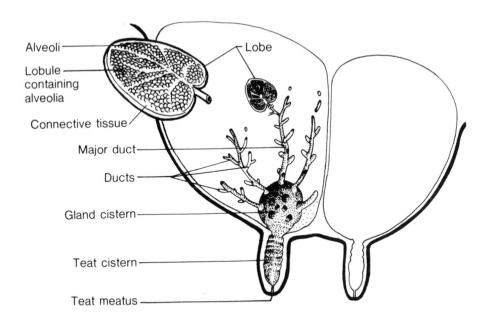

Alveoli
Lobule containing alveolia
Connective tissue
Major duct
Ducts
Gland cistern
Teat cistern
Teat meatus
Lobe

FIGURE 10-1. Diagram of the duct system in one quarter of the mammary gland of the cow with a single lobe illustrated. Four quarters are fused into a single gland complex.

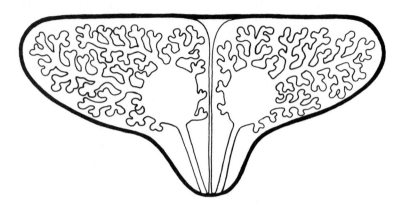

FIGURE 10-2.   Diagram of the gland complex found in the mare.

Lactation has been studied more extensively in cows than in other species. Therefore, the cow will be used as a model in discussing both morphology and function of the mammary gland.

**10-1.2**
**Morphology**

The mammary gland can be divided into supporting tissues and those tissues involved in synthesis and transport of milk. The supporting structures are skin, ligaments and connective tissue. The major support comes from *lateral suspensory ligaments*, which are not elastic, and the *median suspensory ligament*, which is elastic (Figure 10-3). The lateral suspensory ligaments are over the outside of the udder just under the skin. In addition to enveloping the udder, the lateral suspensory ligament sends *lamellae* (thin, convex layers of connective tissue) into the udder. These lamellae become continuous with the interstitial framework of the udder adding to its support. The median suspensory ligament forms longitudinally between the two halves of the udder and is fused on the abdomen. Since it is elastic, it will stretch as the udder fills with milk. Lamellae are given off by the median suspensory ligament, also. The role of skin in support of mammary glands is small when compared with lateral suspensory ligaments and the median suspensory ligament.

Connective tissue divides the milk synthesis and transport system into many subdivisions. The larger of these subdivisions is the *lobe*. The lobe is further divided into small subdivisions called *lobules*. Lobules are not further divided by connective tissue and each has one drainage duct. Each lobule contains from 150 to 225 *alveoli*. Alveoli are tiny sac-like structures which are spherical in shape (Figure 10-4). They have a lumen and are lined with epithelial cells. These epithelial cells are the basic milk secretion units in the mammary gland. Over half of all of the milk stored in a mammary gland will be stored in the lumen of alveoli. The rest will be stored in the ducts leading from the lobules and lobes (Figure 10-1).

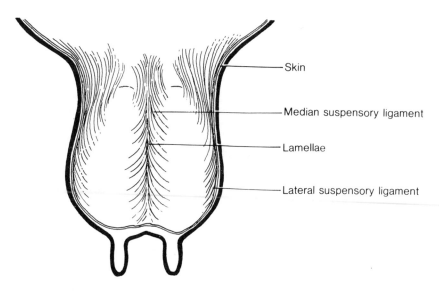

FIGURE 10-3. Diagram of a cross section of the supporting structures of the mammary glands of the cow as viewed from the rear.

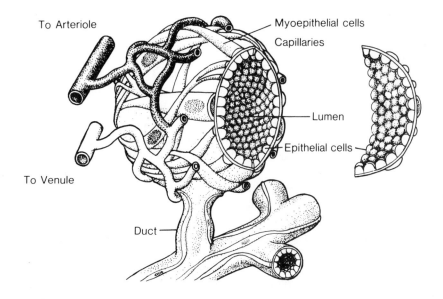

FIGURE 10-4. Diagram of alveolus showing lumen, epithelial cells, myoepithelial cells, and capillaries.

An adequate blood supply to the mammary gland is necessary for the production of milk. Nutrients which are utilized in the synthesis of milk come from blood. Approximately 400 volumes of blood must pass through the mammary gland for the synthesis of one volume of milk. The primary blood supply for mammary glands in cows, mares, sheep and goats is from the *external pudic* artery. In sows, mammary glands are supplied by both external pudic and *external thoracic arteries*. Arteries which penetrate mammary glands branch and follow the connective tissue septa which forms lobes and lobules. Alveoli will be surrounded by a fine network of arterial capillaries where the transfer of nutrients used in milk synthesis takes place. The venous return from the mammary gland is through both the external pudic and *subcutaneous abdominal* veins in those species with mammary glands in the inguinal region. In sows, return is through external pudic and external thoracic veins.

A network of *myoepithelial cells* covers the surfaces of alveoli and small ducts down as far as ducts draining the lobules. They have a smooth musclelike function, but are of ectodermal rather than mesodermal origin. They originate from epithelial cells. Myoepithelial cells are the contractile tissues which play a major role in milk ejection or the "letdown" of milk.

*Smooth muscle fibers* are found in mammary glands. They regulate the size of small arteries and veins, thus controlling the blood supply to the secretory cells. They appear to be too sparse and too irregular to play a major role in the letdown of milk.

**10-2**
**Hormonal regulation of the development and function of the mammary gland**

**10-2.1**
Mammary development

Development of the mammary gland can be divided into four phases. These are embryonic development, fetal development, postnatal growth period development, and development during pregnancy. Differences occur in the regulation of each phase. The first evidence of mammary development in the embryo is the mammary band, a small thickened area of epithelial cells that is seen at about 30 days in cattle. The mammary gland is of ectodermal origin. The developmental stages that follow are the *mammary line, mammary crest, mammary hillock,* and *mammary bud*. Mammary buds can be seen by the early part of the "period of the fetus." There is little evidence that the embryonic development of the mammary gland is under endocrine control. Mammary buds are seen in both males and females.

A *primary sprout* will form in the mammary tissue of the fetus by the third month of gestation. The primary sprout is the beginning of the milk secretion tissue that will form. Before the end of gestation, *secondary sprouts* and possibly *tertiary sprouts* will have formed. While regulation of this phase in development is not completely understood, there is evidence for endocrine influence. Prolactin in synergism with insulin, steroid hormones of the adrenal cortex, and possibly progesterone stimulates this development.

After birth, mammary development will continue in nonpregnant heif-, ers up to about 30 months of age. The major development noted will be the replacement of fatty tissues in the mammary glands with ductal tissue. A surge in the development of ductal tissue will start approximately three months before puberty and continue for several months after puberty. During the mammary growth surge, the rate of growth of mammary tissue will approach 3.5 times that of body growth. The rate of development after 12 months of age will decline to close to the body growth rate. The ductal growth that occurs in a nonpregnant heifer is believed to be due to cyclic surges in estrogen that start several months before puberty and continue with each estrous cycle after puberty.

When pregnancy occurs, mammary growth will continue throughout the pregnancy. Estrogens are dominant in stimulating development of mammary ducts. Progesterone synergizes with estrogens and appears dominant in stimulating alveolar development and growth. Other hormones which synergize with estrogens and progesterone in preparing mammary tissue for secretion of milk include prolactin, growth hormone, insulin, thyroid hormone, and cortisol. Placental lactogen is produced by the placenta and has been identified in some mammals. It will stimulate the development of mammary tissue, but its interaction with other hormones involved in this process is not understood. It should be noted that while mammary tissue is prepared for milk synthesis during gestation, actual secretion of milk is inhibited until just before parturition. The high concentration of progesterone, which is seen during most of the gestation, may be the cause.

**10-2.2
Milk secretion**

Hormone changes seen during late gestation (see Chapter 9) not only initiate parturition, but also initiate milk production. Prolactin is the dominant hormone in initiating lactation in most species that have been studied. The nursing stimulus, as well as many other stimuli, will trigger the release of prolactin. The nursing stimulus along with milk removal is probably more important to the maintenance of lactation than other stimuli. While prolactin is dominant, it must interact with other hormones to attain its greatest effect. Hormones which synergize with prolactin in stimulating lactation are cortisol, growth hormone, thyroid hormone, and insulin. In cows, growth hormone is more dominant than prolactin in maintaining lactation after peak milk production is reached, approximately 2 months into the lactation. As with prolactin, growth hormone is released by the nursing stimulus.

An adequate level of nutrition is necessary for the maintenance of lactation. However, during the first few weeks after parturition, lactation will proceed at the expense of the body reserves of the mother. This provides needed nutrients for the young but may extend postpartum anestrus well beyond the desired interval (see Chapter 9).

10-2.3
Milk ejection

Milk ejection, often called "letdown" of milk, is physiologically a separate function from milk synthesis. Milk ejection is triggered by stimulation of sensory nerves in the teats, by either suckling the young or having the teats massaged (Figure 10-5). This stimulation results in the release of oxytocin from the posterior pituitary. Oxytocin reaches the mammary gland by way of arterial circulation. It stimulates the myoepithelial cells surrounding alveoli and small ducts, forcing the milk down into larger ducts, gland cisterns, and teats where it can be readily removed. While stimulation of sensory nerves in teats will trigger this milk ejection reflex, milk ejection can become a conditioned response. The presence of young, even though physically separated from their mother, will sometimes cause milk ejection. A similar re-

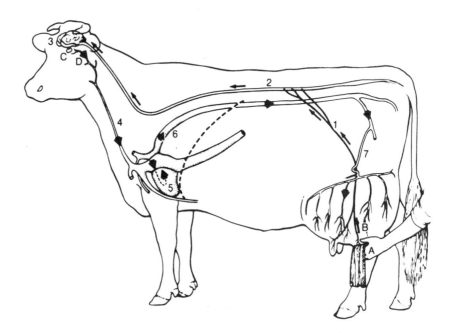

FIGURE 10-5. The neurohormonal reflex of milk ejection. The stimulus (A) that a cow associates with milking causes a nerve impulse (B) to travel via the inguinal nerve (1) to the spinal cord (2) and the brain (3). The brain causes the release of oxytocin (D) from the posterior pituitary (C). Oxytocin is released into a branch of the jugular vein (4) and travels to the heart (5) and is then transported to all parts of the body by the arterial blood. The oxytocin reaching the udder leaves the heart by the aorta (6) and enters the udder through the external pudic arteries (7). In the udder, it causes the myoepithelial cells to contract, resulting in milk ejection from the alveoli. (Schmidt. 1971. *Biology of Lactation*. W. H. Freeman and Co. Redrawn by permission of the author.)

sponse is sometimes seen in cows waiting to be milked in a milking parlor. Sounds or odors associated with milking may trigger the release of oxytocin.

Excitement, resulting in the release of epinephrine into the blood stream, will inhibit the milk ejection reflex. One effect from epinephrine is vasoconstrictions of small arteries and veins. Vasoconstriction of small arteries may prevent oxytocin from reaching the myoepithelial cells. Another effect from epinephrine might be inhibition of the release of oxytocin from the posterior pituitary.

**10-3
Composition
of milk**

Even though milk is the sole nourishment for neonatal mammals, its composition varies among species. Fat is the most variable component ranging from 1.3% in donkeys to 30 to 50% in sea mammals. Even in those species whose milk is sometimes used for human food, milk fat ranges from approximately 2% in the mare to 11% in ewes (Table 10-2). Breed differences in milk composition are seen also. Again fat is the most variable component with averages ranging from 3.6% in Holsteins to 4.9% in Jerseys.

TABLE 10-2. Species and breed differences in milk composition

| Species | Fat | Protein | Lactose | Ash |
|---|---|---|---|---|
| Horse | 1.6 | 2.7 | 6.1 | 0.5 |
| Sheep | 11.0 | 7.0 | 3.5 | 0.9 |
| Swine | 7.5 | 5.4 | 4.7 | 0.9 |
| Goat | 7.9 | 6.1 | 3.4 | 0.9 |
| Cattle | | | | |
| Guernsey | 4.7 | 3.6 | 4.8 | 0.7 |
| Holstein | 3.7 | 3.1 | 4.6 | 0.7 |
| Jersey | 4.9 | 3.7 | 4.8 | 0.7 |
| Shorthorn | 3.6 | 3.3 | 4.5 | 0.8 |

Marked differences are seen when comparing colostrum, the first milk secreted by the mammary gland near parturition, with normal milk that is produced through the rest of the lactation (Table 10-3). Most notable when comparing colostrum with normal milk is the much higher protein level in colostrum. Colostrum is especially higher in the gamma globulin fraction of protein which is the immunoglobulin or antibody portion of milk. Colostrum is higher in Vitamin A than is normal milk. The newborn is deficient in this vitamin. Colostrum is also higher in minerals.

Composition of milk during the lactation varies over a narrow range after the transition from colostrum to normal milk. During late lactation when a smaller volume is being produced, the fat percentage will be slightly higher. During milk removal, the first milk removed will be relatively low in fat, whereas that obtained by stripping the last of the milk from the mammary gland will be high in fat. Therefore, to get a true picture of its composition, milk should be removed from the gland as completely as possible and mixed before sampling.

TABLE 10-3.  Comparison of the composition of colostrum with normal milk

| | Cow | | Sow | | Mare | |
|---|---|---|---|---|---|---|
| Constituent | Colostrum | Milk | Colostrum | Milk | Colostrum | Milk |
| Total Solids (%) | 23.9 | 12.9 | 20.5 | 16.9 | 25.2 | 11.3 |
| Fat (%) | 6.7 | -4.0 | 5.8 | 5.4 | 0.7 | 2.0 |
| Protein (%) | 14.0 | 3.1 | 10.6 | 5.1 | 19.1 | 2.7 |
| Lactose (%) | 2.7 | 5.0 | 3.4 | 5.7 | 4.6 | 6.1 |
| Ash (%) | 1.11 | 0.74 | 0.73 | 0.71 | 0.72 | 0.50 |

From G. H. Schmidt, *Biology of Lactation*, W. H. Freeman, 1971. By permission of author.

**Suggested reading**

Baldwin, R. L. and T. Plucinski. 1977. Mammary gland development and lactation. *Reproduction in Domestic Animals.* (3rd ed.) ed. H. H. Cole and P. T. Cupps. Academic Press.

Cowie, A. T. 1972. Lactation and its hormonal control. *Reproduction in Mammals. 3. Hormones of Reproduction.* ed. C. R. Austin and R. V. Short. Cambridge Press.

Cowie, A. T. and J. S. Tindal. 1971. *The Physiology of Lactation.* Monographs of the Physiological Society. Edward Arnold. London.

Larson, B. L. and V. R. Smith. 1974. *Lactation: A Comprehensive Treatise.* Vols. I, II and III. Academic Press.

Schmidt, G. H. 1971. *Biology of Lactation.* W. H. Freeman and Co.

# Chapter

# 11

# Male mating behavior

The primary purpose of mating behavior is copulation, thus bringing male and female gametes together to insure propagation of the species. For males the events in mating behavior, listed sequentially, are sexual arousal, courtship (sexual display), erection, mounting, intromission (insertion of penis), ejaculation, and dismounting. If the female is in estrus, this sequence of events takes only a few minutes in cattle and sheep. In horses and pigs, the duration of courtship and copulation is extended, with copulation alone lasting for 10 to 20 minutes in pigs.

Both similarities and differences exist in the courtship patterns of different species (Figure 11-1). Vocalization occurs in most species. This may be bellowing by the bull, neighing by the stallion, or grunts by rams and boars. Sniffing of the female's genitalia and urine is seen. In cattle, sheep, and horses the male will extend his neck and curl his upper lip. Various tactile stimuli to include licking, bunting, and biting are part of the courtship pattern in most males. In addition, the male will try to protect the female, separating her from other males and females. Thus, dominant males prevent subordinate males from copulating.

## 11-1 Regulation of mating behavior

### 11-1.1 Hormonal influence

Increased aggressiveness and sexual desire can be correlated with increased production of testosterone as a male approaches puberty. In addition, the reduction in sexual activity associated with castration can be restored by injections or infusions of testosterone. Seasonal reductions of sexual activity in stallions, rams, and bucks (goat) have been associated with seasonal reductions of testosterone. Such seasonal phenomena appear to be regulated by daylength patterns, as in the female. Reductions in testosterone and sexual activity during the summer in other species (cattle, pigs) are probably related to reduced thyroid activity, resulting in lower basal metabolic rates rather than to a direct effect of summer temperatures on the Leydig cells. These examples illustrate the dominance of testosterone in

(a) Ram sniffing vulva of the ewe.

(b) Stallion nuzzles mare's head and paws ground.

(c) Boar and sow engage in mating grunts.

(d) Bull showing curled upper lip.

FIGURE 11-1. Courtship in farm animals.

regulating male sexual behavior. It should be noted that testosterone will increase sexual activity to a certain threshold, after which higher levels of testosterone will not accentuate the response. For example, in castrated males, infusion of testosterone will restore sexual activity to its previous level, but will not elicit greater activity than was seen before castration.

While testosterone has a major controlling influence on male sexual behavior, it interacts with other factors in eliciting the full response. If males are castrated before puberty, sexual activity is greatly diminished, but some mounting behavior may continue. If sexually experienced males are castrated, they may continue to copulate for several years. However, the frequency of this activity will be reduced, and the ability to ejaculate is probably lost with involution of the accessory glands. It is not clear if the continued sexual activity of castrated males is due to physiological or psychological mechanisms, but certain other factors have been shown to influence mating behavior in intact (not castrated) males (Figure 11-2).

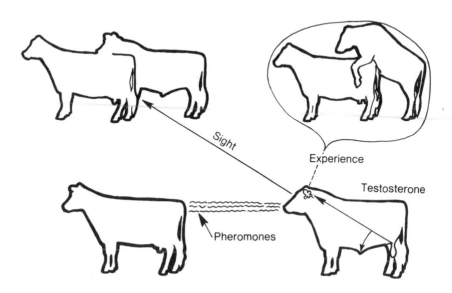

FIGURE 11-2. Interaction of factors that regulate sexual behavior in bulls and other species.

**11-1.2 Social and sexual interaction**

Social experience during development before puberty is important to the attainment of full sexual activity. It has been demonstrated that males raised in complete isolation never develop as high a level of sexual activity as males raised in social groups. The sex of the animals in the social group does not matter. Social interaction with others of the same species is the key to attaining full sexual activity.

Lack of early sexual experience has resulted in neither serious nor permanent problems in attaining full sexual activity. In artificial insemination centers, sexually inexperienced bulls are sometimes hesitant, spend a long time exploring the genitalia of female teaser animals, and frequently have weak erections with incomplete ejaculation. However, on repeated use under the same conditions, they adjust to the surroundings and copulate vigorously.

**11-1.3 Senses**

Certain senses are important to the mating response of males. The sense of smell may be the most important, as in females. Pheromones found in the urine of estrous females are stimulatory to males. Both bulls and stallions can be trained to mount dummies and serve artificial vaginas when collecting semen for use in artificial insemination. When these males do not respond to the dummy mount, sprinkling urine from an estrous female on the mount will frequently elicit the desired response. It was noted earlier in this chapter that sniffing the vulva and urine of females in estrus is a part of the courtship of most farm species.

The sense of sight is probably more important to the mating response of males than of females. The importance of this sense appears to relate to sexual arousal rather than to other aspects of mating. Sight of an estrous female, a castrate, a padded dummy mount, or any object that can be mounted will usually stimulate mounting behavior in bulls. Sexually mature bulls wearing blindfolds will copulate normally. On the other hand, males blind from birth reach sexual maturity at a later date than do normal males.

The tactile sense (sense of touch) is important to normal copulatory behavior in most males. Males may be sensitive to both temperature and pressure. Both bulls and rams are sensitive to the temperature of the water in the artificial vagina (see Chapter 14). Most bulls respond best to the artificial vagina if the temperature is approximately 45°C. If the temperature of the artificial vagina is too hot or too cold, he will not ejaculate and may become uncooperative in future attempts to collect semen. Boars and stallions are more sensitive to pressure on the penis than to temperature. For example, a boar will not ejaculate unless pressure is applied to his glans penis. This pressure may be from the hand of an attendant or the cervix of a sow as the boar's penis locks into the cervix during copulation.

The importance of the other senses has not been clearly demonstrated. The sense of hearing is apparently less important than in females. The sense of taste may be important through its relationship to the sense of smell.

The effects of all of the senses are additive. The response of one is accentuated by an additional stimulus. Also, if one sense is eliminated, the male may make more use of other senses and perform normally. Senses interact with both experience (social and sexual) and conditioned reflexes. They can not be completely separated.

**11-2**
**Erection and ejaculation**

**11-2.1**
Erection

Erection is under control of the autonomic nervous system. With sexual excitement, blood is pumped into and temporarily trapped in the corpus cavernosum penis and corpus spongiosum penis. The penis of the stallion has large cavernous areas which account for the increase in size of the penis during erection. In the bull, ram, and boar erection results in extension of the penis with little increase in size. They have fibroelastic penises with small areas of cavernous tissue. The corpus cavernosum penis is larger and more important in attaining an erection than the corpus spongiosum penis. Pressure in the corpus cavernosum penis just before ejaculation may exceed 15,000 mm mercury in bulls and 6500 mm mercury in stallions. The energy for this pressure comes from the ischiocavernosus muscle, which contracts to pump blood into and traps blood in the corpus cavernosum penis. The corpus cavernosum penis is a closed system with no venous outlet to bleed off the pressure.

**11-2.2
Ejaculation**

Ejaculation is defined as the ejection of semen from the body. The ejaculate includes spermatozoa from the vas deferens and epididymis and fluids from the accessory glands. Ejaculation is initiated by stimulation of sensory nerves in the glans penis, which triggers a series of peristaltic contractions involving smooth muscles in the epididymis, vas deferens, and urethra. In addition, fluids from the accessory glands are pumped into the urethra. Peristaltic contractions move spermatozoa concentrate and accessory gland fluids through the ducts leading to the external urethral orifice. The final discharge of semen is brought about by a wave of contractions involving the smooth muscles lining the urethra and in bulls by pressure from the corpus spongiosum penis which collapses the urethra in a wave. The pressure from the corpus spongiosum penis is generated by the bulbospongiosum muscle and proceeds from the penile bulb towards the glans penis.

Ejaculation varies among species in a number of aspects (Table 11-1). Ejaculation occurs almost instantaneously with the thrust of the penis in bulls and rams. Ejaculation time is about 10 to 15 seconds in stallions and 10 to 20 minutes in boars. Volumes of ejaculates vary considerably, being smaller in rams (1.5cc) and bulls (7cc) and much larger in stallions (75cc) and boars (300cc). Concentration of spermatozoa ranges from a low of approximately 150 million per cc in stallions to a high of 2 billion per cc in rams. Boars (200 million per cc) and bulls (1.2 billion per cc) are intermediate in spermatozoa concentration. The consistency of the ejaculate varies among species. In bulls and rams, there is a complete mixing of spermatozoa concentrate with fluids from the accessory glands in the urethra before expulsion of the semen. Dribblings from the prepuce of the bull before copulation, thought to be from the bulbourethral glands, are quite low in volume. Both boars and stallions have a segmented ejaculate. A spermatozoa-free segment will be followed by a spermatozoa-rich segment and a spermatozoa-poor

TABLE 11-1.   Characteristics of average ejaculate of semen for different species

| Species | Ejaculation time | Volume (cc) | Concentration |
|---------|------------------|-------------|---------------|
| Bull | Less than 1 sec | 7 | 1.2 billion/cc |
| Ram | Less than 1 sec | 1.5 | 2.0 billion/cc |
| Boar | 10-20 minutes | 300 | 200 million/cc |
| Stallion | 10-15 sec | 75 | 150 million/cc |

segment. When collecting boar semen for artificial insemination, the initial spermatozoa-free segment can be discarded without influencing fertility (see Chapter 14).

**11-3 Maintaining libido**

Maintaining *libido* (sex drive) in males is of concern in both range (or farm) conditions and with artificial insemination organizations. The problem has been studied more extensively in bulls than in other farm species.

Providing a balanced diet is highly important. This means to neither underfeed nor overfeed. As mentioned in Chapter 6, underfeeding will delay puberty. It has also been demonstrated that underfeeding young bulls between the age of puberty (10 to 12 months) and the end of their growth period (30 to 36 months), will reduce libido and total semen production. In mature bulls overfeeding is more likely to be a problem. If mature bulls are overweight, the increased likelihood of foot, leg, and joint problems may shorten their reproductive life. Adequate but restricted diets do not reduce libido or semen production in mature bulls.

Diseases and injuries will reduce libido. Neither of these conditions has to be serious to reduce sexual activity. A seemingly minor problem, such as a swollen joint or mild respiratory infection, may reduce the overall vigor of the bull as well as his libido. When bulls or other males are confined to small areas, special attention must be given to hoof care (trimming, etc.) and moderate exercise to help maintain a state of good health and prevent conditions that make copulation unpleasant. If copulation is unpleasant or causes pain, voluntary retirement from sexual activity is not uncommon.

**11-3.1 Sexual exhaustion**

Sexual exhaustion can be a problem when males run with females for a prescribed breeding season. A primary effect of sexual exhaustion is loss of libido. This is more likely to happen with a dominant male during the heavy part of the breeding season under range conditions. Subsequent gestations have been delayed because dominant, but sexually exhausted males have prevented subordinate males from copulating. If a number of females are in estrus, or if at least two other males are running with the herd or flock, the dominant but sexually exhausted male will be less able to prevent breeding of estrous females. After adequate rest, males that have been sexually exhausted will regain their libido.

It has been reported that excessive copulation by a stallion will reduce fertility and in some cases cause sterility. However, mature stallions in good health have been used daily and sometimes more frequently for extended periods without loss of fertility. Reduced libido occurs before reduced fertility in other species and this may be true in stallions also.

**11-3.2 Sexual satiety**

Whereas sexual exhaustion is a physical problem that can be cured with rest, sexual satiety is a mental problem. It is a term used to describe a sexual indifference that results when a bull has copulated with the same cow under the same conditions for an extended period. It is a problem which has been encountered by organizations where bulls are maintained exclusively for the purpose of providing semen for artificial insemination. In most cases sexual satiety can be cured by providing variety in the collection procedure. This might be done by changing the teaser animal, bringing out a second teaser, or moving to another collection area. Use of an estrous female or sprinkling the collection area with urine from an estrous female will frequently help.

Technicians who collect semen from bulls using an artificial vagina must be careful to make the experience pleasant for the bull. Having the water in the artificial vagina too hot or too cold, excessive bending of the penis, or grasping the penis will be unpleasant and may make the bull uncooperative on future collection attempts. (See Chapter 14 for proper semen collection procedures.)

**Suggested reading**

Alexander, G., J. P. Signoret and E. S. E. Hafez. 1974. Sexual and maternal behavior. *Reproduction in Farm Animals*. (3rd ed.) ed. E. S. E. Hafez. Lea and Febiger.

Beckett, S. D., R. S. Hudson, D. F. Walker, R. I. Vachon and T. M. Reynolds. 1972. Corpus cavernosum penis pressure and external penile muscle activity during erection in the goat. *Biol. Reprod.*, 7:359.

Hafs, H. D. and M. S. McCarthy 1979. Hormonal control of testis function. *Beltsville Symposia in Agricultural Research. 3. Animal Reproduction*. ed. H. W. Hawk. Allanheld, Osmun and Co., Halsted Press, Wiley & Sons.

Hale, E. B. and J. O. Almquist. 1960. Relation of sexual behavior to germ cell output in farm animals. 4th Biennial Symp. on Animal Reproduction. *J. Dairy Sci.*, 43 (supp):145.

Hulet, C. V., S. K. Ercanbrack, R. L. Blackwell, D. A. Price and L. O. Wilson. 1962. Mating behavior of the ram in the multi-sire pen. *J. Anim. Sci.*, 21:865.

Katongole, C. B., F. Naftolin and R. V. Short. 1971. Relationship between blood levels of luteinizing hormone and testosterone in bulls and the effects of sexual stimulation. *J. Endocrin.*, 50:458.

Lindsey, D. R. 1965. The importance of olfactory stimuli in the mating behavior of the ram. *Animal Behavior*, 13:75.

Melrose, D. R., H. C. B. Reed and R. L. S. Patterson. 1971. Androgen steroids associated with boar odour as an aid to the detection of oestrus in pig artificial insemination. *Brit. Vet. J.*, 137:497.

Seidel, G. E., Jr. and R. H. Foote. 1969. Motion picture analysis of ejaculation in the bull. *J. Reprod. Fert.*, 20:317.

Wierzbowski, S. 1966. The scheme of sexual behavior in bulls, rams and stallions. *World Rev. Animal Prod.*, 2:66.

# Chapter

# 12 Semen and its components

Semen is composed of spermatozoa and seminal plasma. Its sources are the epididymis and vas deferens which supply the cellular components (spermatozoa) and the accessory glands which provide most of the fluid portion (seminal plasma). The relative contribution, on a volume basis, is illustrated in Table 12-1 for the bull and boar. In terms of total volume, the contribution of the epididymis and vas deferens is relatively small. In bulls the greatest contribution to the fluid volume of semen is from the vesicular glands, with minor contributions from the prostate gland and bulbourethral glands. In boars, there are greater contributions from the prostate and bulbourethral glands with a smaller proportion from the vesicular glands. These differences are reflected in the chemical composition of semen, also (Table 12-2). Bull semen is higher in fructose and sorbitol, which comes from the vesicular glands, whereas boar semen is higher in most minerals, the major source of these being the prostate gland.

TABLE 12-1.  Sources and relative contribution (volume %) to semen

| | Sources | | | |
| Species | Epididymis and vas deferens | Vesicular glands | Prostate gland | Bulbourethral glands |
|---|---|---|---|---|
| Bull | 5-15 | 60-80 | 10 | 5 |
| Boar | 2-5 | 15-20 | 50-75 | 10-25 |

**12-1**
**Spermatozoa**

The concentration (no./cc) of spermatozoa in an ejaculate of semen is approximately 150 million for stallions, 200 million for boars, 1.2 billion for bulls, and 2 billion for rams (Table 12-3). Theoretically 50% of the sper-

TABLE 12-2.   Average chemical composition of semen from different species (mg/100 cc)

| Constituent | Bull | Ram | Boar | Stallion |
|---|---|---|---|---|
| Fructose | 530 | 250 | 13 | 2 |
| Sorbitol | 75 | 72 | 12 | 40 |
| Glycerylphosphorylcholine | 350 | 1650 | 175 | 70 |
| Inositol | 35 | 12 | 530 | 30 |
| Citric Acid | 720 | 140 | 130 | 26 |
| Ergothionine | 0 | 0 | 15 | 75 |
| Plasmalogen | 60 | 380 | — | — |
| Sodium | 230 | 190 | 650 | 70 |
| Potassium | 140 | 90 | 240 | 60 |
| Chlorine | 180 | 86 | 330 | 270 |
| Calcium | 44 | 11 | 5 | 20 |
| Magnesium | 9 | 8 | 11 | 3 |

Adapted from White, *Reproduction in Farm Animals*, (3rd ed.) ed. Hafez, Lea & Febiger, 1974.


TABLE 12-3.   Characteristics of semen from farm animals

| | Cattle | | | | |
|---|---|---|---|---|---|
| | Dairy | Beef | Sheep | Swine | Horses |
| Volume (cc) | 6 | 4 | 1 | 225* | 60* |
| Sperm concentration (billion/cc) | 1.2 | 1.0 | 3.0 | 0.2 | 0.15 |
| Total sperm (billion) | 7 | 4 | 3 | 45 | 9 |
| Motile sperm (%) | 70 | 65 | 75 | 60 | 70 |
| Morphologically Normal sperm (%) | 80 | 80 | 90 | 60 | 70 |
| pH | 6.5-7.0 | 6.5-7.0 | 5.9-7.3 | 6.8-7.5 | 6.2-7.8 |

*Gel free portion
Adapted from Setchell, *Reproduction in Domestic Animals*, (3rd ed). ed. Cole and Cupps, Academic Press, 1977.

matozoa in a given ejaculate will contain X chromosomes and 50% Y chromosomes, which on a population basis would result in equal numbers of male and female offspring. Approximately 60 to 70% of the spermatozoa in semen are expected to be progressively motile with an average speed of 6 mm per minute. In high quality semen, 80 to 90% of the spermatozoa will have normal morphology. Concentration, motility percent, and morphology are all important criteria in the evaluation of semen before use in artificial insemination (Chapter 16). The spermatozoa of all farm animals have an overall length of 60 to 70$\mu$. The head is 8 to 10 $\mu$ long with the tail accounting for the remainder. The head is flattened, about 4 $\mu$ wide and 1 $\mu$ thick.

**12-1.1**
**Normal**
**morphology**

The normal spermatozoon is composed of a head and a tail that is divided into a mid-piece, main-piece and end-piece (Figure 12-1).

The important components of the head include the *nucleus* containing the genetic code, which is the sire's contribution to a new offspring, the *postnuclear cap*, covering the posterior portion of the nucleus, and the *acrosome*. The acrosome covers the anterior part of the nucleus and contains enzymes needed for penetration of the corona radiata and zona pellucida during fertilization (Chapter 7). If the acrosome is malformed, damaged, or missing the spermatozoon will not be able to participate in fertilization. During aging the acrosome becomes loosened from the nucleus starting at the apical ridge.

The point where the tail joins the head contains the *proximal centriole*, and is called the implantation region. The head and tail become separated at this point during fertilization. Similar separation is sometimes seen in heat-damaged semen.

The mid-piece, a thickened portion of the tail some 8 to 10 $\mu$ long, is located just posterior to the proximal centriole. The *mitochondrial sheath*, which forms from the mitochondria of the spermatid, is a part of the mid-piece. The mitochondrial sheath contains enzymes which convert fructose and other energy substrates into high-energy compounds that can be used by spermatozoa.

The main-piece (40 to 50 $\mu$ long) and end-piece (3 $\mu$ long) differ in that the end-piece does not have a protective sheath. A major feature of the tail is the *axial filament*. The axial filament is a small bundle of tiny fibrils that starts at the proximal centriole and runs through the entire tail. One center pair of small fibrils is surrounded by a circle of nine pair of small fibrils (Figure 12-1). Nine larger fibrils surround the circle of nine pair of small fibrils through much of the length of the tail. Contractions of these fibrils cause a lashing of the tail which propels the spermatozoon forward. Contractions start at the proximal centriole proceeding sequentially around the perimeter fibrils and rhythmically down the tail. This results in an urn-shaped pattern of tail movement causing a rotation of the entire spermatozoon as it moves progressively forward. This progressive motility can be observed under a light microscope.

**12-1.2**
**Abnormal**
**morphology**

Every ejaculate of semen will contain some morphologically abnormal spermatozoa. The expected range of 8 to 10% has no adverse effect on fertility. If abnormal spermatozoa exceed 25% of the total in an ejaculate, reduced fertility can be anticipated.

Abnormal spermatozoa can be classified under abnormal heads, abnor-

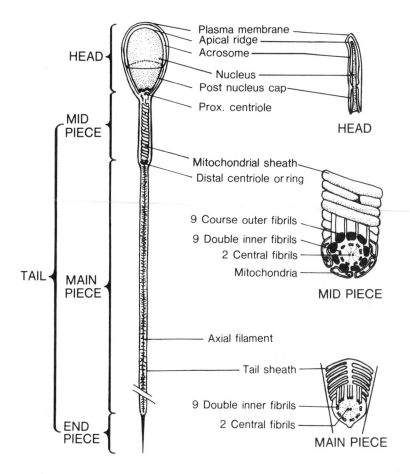

HEAD {
- Plasma membrane
- Apical ridge
- Acrosome
- Nucleus
- Post nucleus cap
- Prox. centriole

HEAD

MID PIECE {

- Mitochondrial sheath
- Distal centriole or ring

- 9 Course outer fibrils
- 9 Double inner fibrils
- 2 Central fibrils
- Mitochondria

MID PIECE

TAIL { MAIN PIECE {

- Axial filament

- Tail sheath

- 9 Double inner fibrils
- 2 Central fibrils

MAIN PIECE

END PIECE {

FIGURE 12-1.  Structural diagram of a spermatozoon. (Adapted from Wu. 1966: *Al Digest.* 14:7.)

mal tails, and cytoplasmic droplets (Figure 12-2). Abnormal heads that have been observed include asymmetrical, tapering, pyriform, giant, micro, and double heads. Abnormal tails include enlarged, broken, bent, filiform, truncated, and double mid-pieces, along with coiled, looped, and double tails. Most spermatozoa with tail abnormalities will not be motile and the remainder exhibit abnormal motility. Cytoplasmic droplets form on the neck of spermatozoa during spermiogenesis. As discussed in Chapter 3, these are usually lost during maturation in the epididymis. If they are still present when spermatozoa are ejaculated they are considered an abnormality and as with other abnormalities, too high a percentage will reduce the fertility of the semen. Stress causes an increase in abnormal sperm. Abnormalities of all types increase but the first to appear and the last to disappear are increases in cytoplasmic droplets.

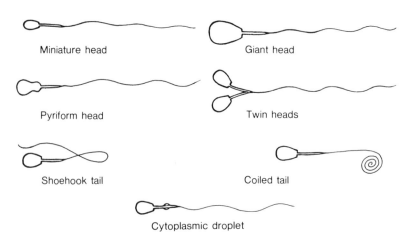

FIGURE 12-2. Morphological abnormalities of spermatozoa identified through examination of semen for quality.

## 12-2
## Seminal
## plasma

The fluid portion of semen is seminal plasma. The accessory glands contribute most of this, but a small amount of fluid is a part of the spermatozoa concentrate which comes from the epididymis and vas deferens. Seminal plasma serves as a buffered nutrient medium which suspends and maintains the fertility of spermatozoa. Seminal plasma is slightly acid in bulls and rams and slightly alkaline in boars and stallions. The osmotic pressure of seminal plasma is similar to blood (equivalent to physiological saline — 0.9% sodium chloride). A number of organic and inorganic compounds are in solution in seminal plasma.

### 12-2.1
### Inorganic ions

Sodium and chlorine are the principal inorganic ions in seminal plasma. Smaller quantities of calcium and magnesium are found also. Potassium, which is present in substantial amounts in whole semen, is more concentrated in spermatozoa than in the fluid suspending the spermatozoa. Thus when spermatozoa are concentrated, as in the epididymis, the potassium-to-sodium ratio is higher. These inorganic ions are important to the viability of spermatozoa, possibly through their effect on the integrity of the sperm cell membrane. Along with the organic molecules in solution in seminal plasma, the inorganic ions help maintain an osmotic pressure that is optimum for the survival of spermatozoa.

### 12-2.2
### Buffering agents

In addition to inorganic ions, organic ions are found in seminal plasma. The principal organic ion is bicarbonate. It is produced by the vesicular glands and functions as a buffering agent, guarding against changes in the pH of semen. Buffers are not found in sufficient quantities to prevent a reduction in pH when semen is maintained in storage. Therefore, good semen diluters must have buffering capacity (Chapter 16).

**12-2.3**
**Energy**
**substrates**

Several organic compounds which serve primarily as energy substrates for spermatozoa are found in seminal plasma. The principal ones are fructose, sorbitol, and glycerylphosphorylcholine (GPC). Fructose (a simple sugar) and sorbitol (a sugar alcohol) are produced by the vesicular glands, whereas GPC is produced in the epididymis. All are unique in that they are not found in substantial quantities elsewhere in the body.

Fructose can be used by spermatozoa as an energy substrate under the anaerobic (oxygenless) conditions of storage and the aerobic (oxygenated) conditions found in the female tract. Sorbitol and GPC can only be utilized aerobically. In addition, GPC must be acted on by an enzyme found in the female tract before it can be utilized. This enzyme splits the choline from the rest of the molecule forming glycerylphosphate, which can be metabolized as an energy substrate. Lactic acid, a byproduct of the anaerobic metabolism of fructose (Section 12-3) builds up in semen that is being stored and theoretically can be used as an energy substrate when placed in aerobic conditions.

Fructose is found in high concentrations in bull and ram semen, but is much lower in both boar and stallion semen. The low concentration of fructose in boar and stallion semen may contribute to the problems of storing semen from these species.

**12-2.4**
**Other organic**
**compounds**

Compounds found in seminal plasma in rather large concentrations but not used as energy substrates are inositol and citric acid. Both are produced by the accessory glands. Ergothionine is found in the semen of boars and stallions. These compounds are not found in substantial amounts elsewhere in the body.

**12-3**
**Energy**
**metabolism by**
**spermatozoa**

Energy metabolism is the means by which spermatozoa convert energy substrates into usable forms of energy. Enzymes for this conversion are in the mitochondrial sheath. The principal energy substrates in semen are fructose, sorbitol, and GPC, which are found in seminal plasma. In addition, plasmalogen, a lipid found within the spermatozoon, is an energy reserve that can be used when other substrates are limited.

Adenosine triphosphate (ATP), a high energy compound, is the form of energy that can be used by spermatozoa. ATP is converted to ADP yielding 7000 calories per mole of energy by the following reaction:

$$ATP + H_2O \rightleftharpoons ADP + Pi + 7000 \text{ calories/mole}$$

If there was no means of regenerating ATP the spermatozoa would not survive due to lack of energy. Energy substrates provide a means by which ATP can be regenerated from ADP plus inorganic phosphorous (Pi). Fructose serves as a good example since it can be utilized anaerobically and aerobically. The anaerobic reaction is as follows:

$$\text{Fructose} \xrightleftharpoons{No\ O_2} 2 \text{ lactic acid} + 2 \text{ ATP (net yield)}$$

Fructose metabolized anaerobically yields a net of 2 ATP or 14,000 calories. This reaction provides energy to maintain the viability of spermatozoa during storage. However, an end product of this metabolism is lactic acid. If steps are not taken to slow metabolism during storage, the buildup of lactic acid will soon lower the pH of the semen, adversely affecting the viability of spermatozoa.

Under aerobic conditions, the metabolism of fructose is:

$$\text{Fructose} \overset{O_2}{\rightleftarrows} CO_2 + H_2O + 38 \text{ ATP (net yield)}$$

When oxygen is present, metabolism of fructose is 19 times more efficient in terms of energy yielded. The net energy from 38 ATP is 266,000 calories. When sufficient oxygen is present, the fructose molecule is metabolized completely to carbon dioxide and water. There is no buildup of lactic acid. In addition, sorbitol, plasmalogen and, if in the female tract, GPC are available for metabolism and regeneration of ATP. Sorbitol and GPC are metabolized through the same biochemical pathways as fructose. Plasmalogen, a lipid rather than a carbohydrate, utilizes different metabolic pathways, but the needed enzymes are in the mitochondrial sheath.

**12-4
Factors
affecting rate
of metabolism**

Rate of metabolism is the rate at which spermatozoa utilize their energy substrates. Under aerobic conditions it can be monitored by measuring oxygen consumption, liberated carbon dioxide, or by methylene blue reduction. Under anaerobic conditions, the rate of reduction of pH or chemical determination of lactic acid buildup and/or fructose disappearance can be used as measures of metabolic rate. Control of metabolic rate is of interest because a reduction in metabolic rate is necessary to extend the storage life of semen. A number of factors contribute to reduced metabolic rate and extended life of spermatozoa in the epididymis (Chapter 3). In the epididymis, spermatozoa may remain fertile for up to 60 days. However, spermatozoa in a fresh ejaculate of semen will be fertile only for a few hours if steps are not taken to reduce their metabolic rate. The measures used must be reversible without injury to spermatozoa if they are to be practical for semen handling.

12-4.1
Temperature

Metabolic rate increases and the life span of spermatozoa decreases as the temperature of the semen rises. When the temperature rises above 50°C, spermatozoa suffer an irreversible loss of motility. If maintained at body temperature spermatozoa will live for only a few hours due to either exhaustion of available energy substrates, drop in pH due to buildup of lactic acid, or a combination of these factors. Reducing the temperature of the semen will slow metabolic rate and extend the fertile life of semen if precautions are taken to protect against *cold shock* and *freeze kill*.

Spermatozoa of all species studied are susceptible to cold shock if they are cooled too quickly. The most obvious indication of cold shock is an irreversible loss of motility. The most critical range for cold shock occurs when the semen temperature is being reduced from 15 to 0°C. Cold shock protection is provided for bull, ram, and stallion semen by cooling slowly after addition of an egg yolk or milk diluter. Both egg yolk and milk contain lecithin and lipoproteins, which give protection against cold shock.

The second problem in reducing metabolic rate to extend the fertile life of semen is freeze kill. Spermatozoa may be killed during the freezing and/or thawing process, apparently by a disruption of the sperm cell membrane. Equilibration of bull semen in a diluter containing glycerol will give adequate protection. Some freeze kill will still occur, but fertility will not be affected if sufficient motile spermatozoa are placed into the semen unit before freezing. While there has been less success in freezing ram and stallion semen, glycerol is beneficial when freezing semen from these species. Excess glycerol is detrimental to boar spermatozoa during freezing.

Reducing the temperature of semen to lower its metabolic rate has been the most useful means of extending the fertile life of semen because it permits restoration of metabolic rate before insemination. Also, greater reduction of metabolic rate can be achieved without reducing fertility. When bull semen is frozen in liquid nitrogen at $-196°C$, its metabolic activity is reduced to less than 0.02% of the metabolic rate at body temperature. Further, fertility can be maintained in the semen for decades.

**12-4.2**
**pH**

A pH of about 7.0 (6.9 to 7.5 for different species) falls in optimum activity range of most of the enzymes in spermatozoa. Therefore, a higher metabolic rate is expected when the pH of semen is maintained near neutrality (7.0). If the pH of semen deviates toward alkalinity or acidity metabolic rate will be reduced. The practicability of altering the pH of semen to extend its life is limited by the narrow range over which pH can be altered without permanently reducing activity. Research in this area has established the importance of diluting semen in a buffered medium that resists changes in pH, so that maximum fertile life of the semen can be maintained.

**12-4.3**
**Osmotic**
**pressure**

Semen maintains maximum metabolic activity when diluted with an isotonic diluter. Either hypotonic or hypertonic diluters will reduce metabolic rate, but neither will extend the life of the semen. The spermatozoon membrane is a semipermeable membrane. Both hypotonic and hypertonic diluters will alter transfer of water through this membrane, disrupting the integrity of the cell. It is very important that only isotonic diluters be used. Spermatozoa remain motile longest when suspended in isotonic media.

**12-4.4**
**Concentration**
**of spermatozoa**

Increasing the concentration of spermatozoa above that found in the normal ejaculate will decrease metabolic rate. Potassium is the principal cation in the sperm cell, whereas sodium is the major cation in seminal plasma. Increasing the cellular concentration will increase the potassium-to-sodium ratio in the semen. Potassium is a natural metabolic inhibitor. Increasing its concentration will reduce the metabolic activity in the semen.

Generally, moderate dilution of semen in a buffered, isotonic medium containing fructose will not greatly alter metabolic rate, but will extend the life of the semen. Dilution such as this is usually done before lowering the temperature of semen. Some caution must be observed. If dilution is excessive ($>$ 1 to 1000), motility and metabolic rate will be depressed.

## 12-4.5
## Hormones

Testosterone and other androgens depress metabolic rate, but those concentrations found in the male system have no permanent effect. Fluids from the female tract increase the metabolic activity of spermatozoa. This is thought to be primarily an effect from estrogens, but other unidentified factors may be involved. The increased metabolic activity in the female tract likely increases motility, which increases the frequency of collisions between spermatozoa and the oocyte in the oviduct.

## 12-4.6
## Gases

Low concentrations of carbon dioxide stimulate aerobic metabolism of spermatozoa. If the partial pressure of carbon dioxide exceeds 5 to 10%, metabolic rate is depressed. Carbon dioxide has been identified as a factor in regulating metabolic rate in the epididymis. Oxygen is necessary for aerobic metabolism. On the other hand, too high a level of oxygen is toxic and will depress metabolic rate. This is not likely to be a factor in the laboratory unless oxygen or air is being bubbled through the semen. Anaerobic metabolism can proceed under nitrogen, hydrogen, or helium gases with no effect on metabolic rate.

## 12-4.7
## Light

Light intensities that are normally found in the laboratory can depress metabolic rate, motility, and fertility in spermatozoa. The harmful effect is observed only if semen is in contact with oxygen. The enzyme catalase will prevent the harmful effect of light, which suggests that light causes a photochemical reaction in the semen that results in the production of hydrogen peroxide. Semen should be protected from light and never exposed to direct sunlight.

## 12-4.8
## Antimicrobial
## agents

Penicillin and dihydrostreptomycin or neomycin are added to semen during processing to control microbial growth. None have a demonstrated effect on metabolic rate. They sometimes increase fertility of semen from low fertility bulls. Also, these antimicrobial agents may extend the fertile life of the semen by controlling microbes, thus sparing energy substrates for spermatozoa.

**Suggested reading**

Flipse, R. J. and W. R. Anderson. 1969. Metabolism of bovine semen. XIX. Products of fructose metabolism by washed spermatozoa. *J. Dairy Sci.*, 52:1070.

Graves, C. N. 1978. Semen and its components. *Physiology of Reproduction and Artificial Insemination of Cattle.* (2nd ed.) ed. G. W. Salisbury, N. L. VanDemark and J. R. Lodge. W. H. Freeman and Co.

Mann, T. 1964. *The Biochemistry of Semen and the Male Reproductive Tract.* Methuen and Co. Ltd. London.

Sullivan, J. J. 1978. Morphology and motility of spermatozoa. *Physiology of Reproduction and Artificial Insemination of Cattle.* (2nd ed.) G. W. Salisbury, N. L. VanDemark and J. R. Lodge. W. H. Freeman and Co.

White, I. G. 1973. Biochemical aspects of spermatozoa and their environment in the male reproductive tract. *J. Reprod. and Fert.* (supp.) 18:225.

White, I. G. 1974. Mammalian semen. *Reproduction in Farm Animals.* (3rd ed.) ed. E. S. E. Hafez. Lea and Febiger.

Part

# 3    Artificial insemination

# Chapter

# 13

# Introduction and history of artificial insemination

**13-1**
**Introduction**

Artificial insemination (AI) is the most valuable management practice available to the cattle producer. The procedure makes efficient use of the generous supply of sperm available from an individual male in a manner that greatly increases genetic progress as well as improving reproductive efficiency in many situations.

Many bulls produce sufficient semen to provide enough sperm for 40,000 breeding units in one year. A bull is usually at least four years old when his genetic merit has been evaluated, and if he lives to be 10 years old can produce up to 300,000 breeding units of semen. Boar semen can be frozen and utilized more efficiently also. Artificial insemination is available only on a limited basis in sheep and horses.

Reproductive efficiency using AI is at least as good as using natural mating when no diseases are present. When certain diseases enter the picture, especially venereal diseases, AI becomes an important factor in their control.

AI is being used in the farm species other than cattle and swine even though techniques for freezing and thawing are not perfected well enough for commercial use. Researchers have made some progress with freezing stallion semen, but little progress with ram semen. Procedures for identifying superior germ plasm in swine, sheep, and horses have not been developed as well as they have for cattle.

**13-2**
**History**

The first reported use of AI, although not documented, was in 1300 by some Arabian horse breeders. Rival chieftains reportedly stole stallion semen from one another to breed their own mares.

The first documented report of successful use of AI was by an Italian physiologist, L. Spallanzani, in 1780. After success with several amphibian animals he decided to experiment with the dog. He used semen at body temperature to inseminate a bitch which had been confined to his own home. Sixty-two days later she gave birth to three pups. In 1782, Spallanzani's experiment was successfully repeated by P. Rossi and a professor named Branchi.

Spallanzani later demonstrated that the fertilizing component of semen could be filtered out of the seminal fluid. The filtered fluid was sterile while the residue remaining on the filter was highly fertile. In 1803 Spallanzani reported that sperm cooled with snow were not killed but rendered motionless until exposed to heat, after which they were motile for several hours.

His investigations stimulated research of the sex cells and the process of fertilization, but no further reports on artificial insemination were recorded until near the end of the century. Everett Milais, a dog breeder, inseminated nineteen bitches between 1884 and 1887, with fifteen becoming pregnant. Walter Heape of England, writing about the work in 1897, concluded that AI was easy and that conception was as good as natural service. He also suggested that one ejaculate could be used to inseminate several bitches and that AI could be a tool to study genetic factors.

Scientists in the U.S.S.R. began studies with farm animals about 1900. E. I. Ivanoff started his work with horses but was the first to successfully artificially inseminate cattle and sheep. Ivanoff's success stimulated sufficient interest to have a physiological section established in the Ministry of Agriculture specifically to study physiology of fertility and to train veterinarians in the techniques of AI. Work with horses was initiated in Japan about 1913.

The first cooperative AI association was formed in Denmark in 1936. Aided by state support, the Danish breeders have continued to be leaders in the percentage of cows bred by AI. Professor E. J. Perry of Rutgers University was one of the pioneers in the United States. He organized the first AI cooperative in this country in 1938, with 102 members, and bred 1050 cows the first year. Professor Perry visited the Denmark stud and patterned the New Jersey stud after it. Several other cooperatives were organized within the next 2 years. Artificial insemination had become of age and was off with a running start.

A number of important discoveries are worth mentioning. Each raised the level of success of AI to a new plateau.

The first artificial vagina was used to collect dog semen by G. Amantea, a professor of human physiology at the University of Rome (Figure 13-1). Amantea began his research on dog sperm in 1914. Russian scientists patterned artificial vaginas suitable for the stallion, bull and ram after Amanteas'. The development of the artificial vagina suitable for the larger species may well be the most important single development in the history of AI, and the artificial vagina still is preferred where bulls, rams, and stallions are collected on a regular basis. The electroejaculator was developed in the late forties. It has been a useful innovation for collecting from reluctant or disabled bulls and rams.

Artificial insemination was recognized by researchers and many breeders as a tremendous asset for genetic progress by the late 1930s. A limitation was that semen had to be used on the day of collection to give good results. When P. H. Phillips and H. A. Lardy, of the University of Wisconsin, discovered a buffered nutrient media for diluting an ejaculate of semen, it was the first step in correcting the problem. They developed a yolk-phosphate diluter which protected sperm during cooling from body temperature to 5°C, provided a source of nutrients for sperm metabolism and prevented pH change. With this diluter, sperm remained viable and capable of fertilizing ova for three to four days. Salisbury and coworkers improved the diluter by substituting sodium citrate for the phosphates used by Phillips and Lardy. The advantage of the yolk-citrate diluter was sperm visibility under the microscope, permitting more accurate determination of motility after dilution.

The problem of spreading reproductive disease still remained. Efforts were made to collect semen from healthy bulls, but several serious outbreaks of reproductive diseases occurred. The most commonly transmitted disease was vibriosis. Following World War II, penicillin became available for use by the livestock industry. J. O. Almquist, Pennsylvania State University, was the first to report the use of this "wonder drug" to control bacterial contaminants in bovine semen. It was adopted by the AI industry almost immediately with marked improvements in conception rate.

Early inseminations were accomplished by simply depositing semen into the vagina of the female. The technique was later refined by using a speculum and a glass inseminating tube. The speculum was placed into the

FIGURE 13-1.   The earliest artificial vagina, invented by Giuseppe Amantea of the University of Rome, was used for the collection of dog semen. (From *The Artificial Insemination of Farm Animals,* 4th rev. ed., edited by Enos J. Perry. Copyright 1945, 1947, 1952, by the Trustees of Rutgers College in New Jersey. Copyright © 1968 by Rutgers University, The State University of New Jersey.)

vagina, and with a light source (first a battery head lamp and later a pen size flashlight) the posterior of the cervix was visible. The inseminating tube was inserted into the opening and the semen deposited approximately 2 cm into the cervix. In 1937 Danish veterinarians developed the recto-vaginal (or cervical fixation) method of insemination. One hand is inserted into the rectum to manipulate the cervix while an insemination tube inserted through the vagina is passed into or through the cervix. The semen can then be deposited into the anterior cervix and/or the body of the uterus. This technique is still used today.

By the late 1940s there were many AI organizations breeding cows all over the country. A fresh supply of semen had to be shipped to the technician or breeder every two or three days, but AI was being used with good results. Two Englishmen were responsible for the next important discovery. A. S. Parkes and C. Polge developed a successful method for freezing and storing sperm at very low temperature. They discovered that glycerol would protect fowl sperm during the freezing and thawing process. At first this method was not successful with mammalian sperm. Eventually, however, they found that allowing the sperm and glycerol mixture to stand overnight before freezing did work. This standing period is now called equilibration time in which the sperm cells absorb some glycerol to replace part of the free water in the cell. The glycerol acts as antifreeze to prevent water cyrstals from forming during freezing. These workers used dry ice as a refrigerant and stored the sperm at $-79°C$.

In 1957 the American Breeders Service pioneered the use of liquid nitrogen as a refrigerant for freezing and storing semen. Large stainless steel vacuum containers were developed by the Lende Corporation. This made it practical to transport semen long distances and store it on the farm. Containers now available need to be replenished with liquid nitrogen only every 60 to 90 days.

The Dairy Records Processing Center of the USDA began compiling and publishing sire summaries in 1961 that helped evaluate the genetic potential of sires. Prior to that date some states and each bull stud had their own procedure of compiling genetic information. The absence of uniformity made it impossible to compare bulls owned by different studs. The USDA sire summaries provide a great deal more information than was previously available and the information is uniform nationwide. Both production and type information is provided.

The introduction of the straw, a small diameter plastic tube, for packaging semen for freezing probably will not be the last chapter in the AI history book, but at present it is the latest major development. Sorensen introduced the use of straws for packaging semen in 1940. Reports of frozen semen in straws by Pares in 1953 and later by Friis Jakobson in 1956, with refinements by Adler in 1959 and 1961, created sufficient interest to keep researchers active.

The Cassous of L'Aigle, France, a father-and-son team, are credited with the development of the straw to practical application in three stages. The first, in 1964, contained 1.2 cc of semen and showed an encouraging improvement in sperm survival when compared to the 1 cc glass ampule. Realizing that the coefficient of freezing surface was the main factor in determining survival, the Cassous changed to a straw with half the original diameter and a capacity of .5 cc. This straw gave excellent results and has been called the "medium straw." It is the straw being used almost exclusively in the United States. The Cassous developed a still smaller straw with a capacity of .25 cc in 1968 and claimed further improvements in the survival of sperm. This straw is called the "mini straw."

AI organizations and researchers in the United States began tests with the straw during the late 1960s. These organizations began switching from the glass ampules to the straw about 1972. Most of the semen produced in the United States is packaged in straws. In addition to providing greater survival of sperm, the straw requires only about one-third the storage space. This has led to the redesigning of liquid nitrogen storage tanks, particular field units that require less nitrogen and have a longer holding time between recharges. Successful freezing of boar semen became a reality in 1975.

## 13-3 Advantages and disadvantages

The advantages of AI far outweigh the disadvantages. The major advantages are:

1. genetic improvement through more accurate evaluation of transmitting ability of males and greater use of superior germ plasm, even allowing its continued use after a bull's or boar's death;

2. control of venereal and other diseases;

3. improved record keeping on farms where used;

4. more economical than natural service when genetic merit is considered;

5. safer by the elimination of dangerous bulls on the farm, especially for the dairy breeds.

There are few disadvantages to AI. Perhaps the most noted one is that livestock managers must spend a great deal of time checking females for estrus. Some special facilities for corraling and inseminating are required and with beef cattle these will need to be more elaborate. Trained personnel are required to perform the technique.

## Suggested reading

Heape, W. 1897. The artificial insemination of mammals and subsequent possible fertilization or impregnation of their ova. *Proc. Roy. Soc. London*, 61:52.

Marden, W. G. R. 1954. New advances in the electro-ejaculation of the bull. *J. Dairy Sci.*, 37:556.

Perry, E. J. editor. 1968. *The Artificial Insemination of Farm Animals.* Rutgers University Press.

Phillips, P. H. and H. A. Lardy. 1940. A yolk-buffer pabulum for the preservation of bull semen. *J. Dairy Sci.,* 23:399.

Spallanzani, L. 1803. *Tracts on the Natural History of Animals and Vegetables.* (Translated title) (2nd ed.) Edinburgh, Creech and Constable.

Walton, A. 1933. *The Technique of Artificial Insemination.* Edinburgh. Imperial Bureau of Animal Genetics.

# Chapter

# 14

# Semen collection

Semen collection is like harvesting any other farm crop. Effective harvest of semen involves obtaining the maximum number of sperm of highest possible quality in each ejaculate. The ultimate objective is to make maximum use of superior sires. This involves proper semen collection procedures used on males that are sexually stimulated and prepared. The initial quality of semen is determined by the male and cannot be improved even with superior handling and processing methods. However, semen quality can be lowered by improper collection and processing techniques. Semen collection is a complex procedure involving coordinated efforts between the animal handler and the collector. See Table 14-1 for common characteristics of ejaculates for the farm species.

**14-1
Facilities
needed for
semen
collection**

There are several essential features that must be considered in designing facilities for collecting semen. The most important consideration is the safety of the handler and the collector. This is especially true where bulls of the dairy breeds are being collected. Safety fences, usually constructed of 3-inch steel pipe with spaces large enough for a man to step through at 8-foot intervals, should be provided (Figure 14-1). The collection area must provide good footing to prevent slipping and injury to the male being collected. This is best provided by an earthen floor in the immediate collection area. Means to restrain the teaser animals to minimize lateral as well as forward movement must be provided. At the same time easy access for semen collection must be maintained.

**14-2
Methods of
semen
collection**

Several methods of collecting semen have been used since the inception of artificial insemination. The early procedures involved taking the semen from the vagina of the naturally mated female with a spoon or by other means. Later a specially designed rubber breeding bag was placed in the vagina of the cow or mare to catch the ejaculate. The massage method, which

145

involved massaging the vesicular glands and ampullae by way of the rectum, was reported as early as 1925. The quantity and quality of semen obtained by this method presented considerable limitations. The concept of the artificial vagina was presented by Amantea about 1914 and was adapted for use on the larger species by Russian scientists about 1933. A practical version of the electroejaculator was introduced in 1948.

The remainder of this section will be devoted to the artificial vagina and the electroejaculation methods of collecting semen. Essentially all semen used for artificial insemination is collected by these two methods.

TABLE 14-1.   Common characteristics of ejaculates for farm species

|  | Volume cc | Motility % | Concentration sperm/cc × 10⁶ |
|---|---|---|---|
| Dairy bulls |  |  |  |
| Small breeds | 5-6 | 50-80 | 1000-1500 |
| Large breeds | 7-8 | 50-80 | 1000-1500 |
| Beef bulls | 4-5 | 40-70 | 1000-1500 |
| Rams | .75-1.2 | 60-80 | 1500-3000 |
| Boars | 200-300 | 50-70 | 100-150 |
| Stallions | 75-150 | 40-70 | 100-150 |

FIGURE 14-1.   Semen collection area for bulls. Note safety fences over which bulls can be managed while protecting the handler. Earthen floor provides good footing for the bulls. (Courtesy of Select Sires, Inc., Plain City, Ohio.)

14-2.1
Sexual stimulation prior to collection with the artificial vagina

Sexual stimulation by exposing males to the normal courtship situations prior to semen collections increases the number of sperm per ejaculate for all species. The effect of sexual stimulation has not been as well defined for the boar, ram, and stallion as it has been for the bull. The remainder of this section will deal with the bull.

There are two reasons for providing adequate sexual stimulation: (1) to insure that the bull will mount and ejaculate in a reasonable period of time; and (2) to insure the collection of the maximum number of sperm with the highest possible quality per ejaculate. Except for the occasional bull with low sex drive, the latter is the most important. Ejaculates with larger volume, higher concentration, and higher motility will have higher fertility in most cases and more breeding units can be prepared from an ejaculate thus reducing the processing time and cost per breeding unit. In addition the lifetime output of sperm by an individual bull is increased.

Sexual stimulation is accomplished by exposing the bull to the teaser animal for several minutes. False mounts, allowing the bull to mount the teaser animal without ejaculation, enhances the degree of sexual excitement. The combination of a false mount, followed by a few minutes of teasing, plus one or two additional false mounts before collection seems to provide adequate preparation for most bulls. A period of 10 to 15 minutes of teasing without false mounts adequately stimulates most bulls. However, care must be taken to insure that the bull is being stimulated, not just standing.

Bulls generally show a reduction in sex drive if handled in the same manner for a prolonged period of time. Some variations which will help maintain a high degree of sexual stimulus are:

1. Introduction of a new teaser animal, perhaps alternating between a cow and another bull
2. Moving teaser animal to a different position at the same location or to a different location in the collection area
3. Presenting two teasers
4. Allowing teaser or another bull to mount the bull to be collected
5. Bringing a new bull into the collection area
6. Collecting semen from the new bull
7. Avoiding distraction

Bulls that are on a high frequency of collection tend to need more variation in stimulation procedure and may take longer to become sexually prepared than bulls on low collection frequencies. Beef bulls tend to exhibit less sex drive than dairy bulls and therefore require more imagination on the part of the handler in getting them prepared for collection.

Sexual stimulus and preparation increase semen volume and sperm concentration of the ejaculates of all bulls. The increase is greater for some bulls than for others. An increase of 30 to 50% can be expected following adequate preparation. Refer to Table 14-2 for some average increases.

**14-2.2**
**Artificial vagina**

The artificial vagina (AV) method of collecting semen is the fastest and most sanitary of the various methods available. The AV provides a good imitation of the natural vagina. There are a number of modifications of the

TABLE 14-2.  Characteristics of ejaculates from bulls with and without sexual stimulation

| | Volume cc | Motile % | Sperm per ejaculate No. × 10⁹ | Motile sperm per ejaculate No. × 10⁹ | Increase % | Usable ejaculates % |
|---|---|---|---|---|---|---|
| * Unrestrained | 3.7 | 53 | 4.55 | 2.48 | | |
| 1 false mount | 4.3 | 56 | 6.19 | 3.62 | 46 | |
| † Unrestrained | 3.7 | 58 | 4.52 | 2.56 | | 75.8 |
| 1 false mount | 4.5 | 63 | 6.41 | 3.98 | 55 | 97.0 |
| 2 false mounts | 4.9 | 61 | 6.37 | 3.89 | 52 | 93.9 |
| # Beef trial I [a] | | | | | | |
| unrestrained | 5.6 | 54 | 11.82 | 6.41 | | |
| 3 false mounts [c] | 6.2 | 63 | 13.65 | 8.36 | 30 | |
| # Beef trial II [b] | | | | | | |
| unrestrained | 4.6 | 60 | 9.93 | 6.06 | | |
| 3 false mounts [c] | 5.2 | 64 | 11.77 | 7.62 | 25 | |
| # Dairy a | | | | | | |
| 1 false mount | 5.0 | 66 | 13.49 | 8.85 | | |
| 3 false mounts [c] | 6.2 | 68 | 18.23 | 12.26 | 38.5 | |

*From Collins and Bratton, J. Dairy Sci. 34:224-227, 1951.
†From Branton et al., J. Dairy Sci. 35:801-807, 1952.
#From Almquist, J. Animal Sci. 36:331.
a. Figures represent total of two successive ejaculations taken on 1 day per week.
b. Figures represent total of two successive ejaculations taken on 2 days per week.
c. Three false mounts before each ejaculate with 2 min. restraint between first and second false mount.

basic design along with different shapes and sizes for different species. The basic design for the artificial vagina used to collect semen from the mature bull is shown in Figure 14-2. It consists of a rigid rubber casing 40 cm in length and 6.4 cm in diameter. The casing is fitted with a rough-textured latex inner liner. The ends of this liner are folded back over the outer ends of the casing and are secured with tight elastic bands. The space between the liner and the casing is filled with warm water to provide suitable temperature and pressure to evoke ejaculation. One end of the unit is fitted with a rubber funnel that is secured to the casing along with the turned back liner. A graduated glass collection tube is attached to the small end of the funnel.

In one version of the AV there is a hole 1 cm in diameter through the outer casing 10 cm from the end. A latex sleeve half the length of the casing is slipped over that end of the casing to cover the opening. Three functions are served:

1.  The sleeve can be rolled back to expose the opening for filling the AV with water and a small amount of air can be blown into the AV to provide additional pressure.

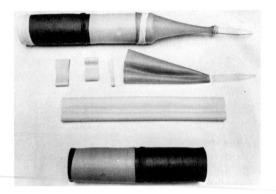

FIGURE 14-2.   Artificial vagina used to collect semen from the bull. The completely assembled AV is shown at the top with the component parts shown below. The protective covering for the funnel and collection tube is shown in Figure 14-5.

2.   A rubber band can be rolled on to the end of the sleeve to hold it tightly, thus providing a chamber to hold displaced water when the penis enters the AV.

3.   This expansion chamber allows the AV to be completely filled with water without danger of the folded inner liner slipping off of the casing.

A loose liner would permit water to escape, possibly ruining the ejaculate of semen. Rubber bands must be counted after each collection to be sure that none remain on the penis.

It is important that the length of the outer casing, and to some degree its diameter, be gauged to fit the males being collected. Young bulls between 10 and 15 months of age should be started with casings 20 to 24 cm long and 5.1 cm in diameter. Casings of intermediate length and diameter need to be used on bulls between 15 months and 3 years of age. For reasons of sanitation it is important that the end of the glans penis extend well into the collecting funnel at the time of ejaculation. Therefore, it is essential that the length of the AV properly correspond to the length of the penis.

The AV can be filled with water through an open end to within 3 to 4 cm of the top while holding the casing vertically. The inner liner is folded back over the end of the casing, being careful not to allow water to spill into the inner liner. The casing may also be filled through the expansion chamber opening (See Figure 14-2). The casing cannot be filled as full using this method. Therefore, to obtain sufficient pressure on the inner liner some air may need to be blown through the opening before the outer sleeve is rolled back over the opening.

The temperature of the water used to fill the AV casing needs to be high enough to bring the internal temperature of the unit to approximately 45°C. Internal temperatures from 38° to 56°C have been used successfully. Most bulls will be stimulated to ejaculate at any temperature within this range. Some will not ejaculate unless the temperature is near 55°C while others may not respond if the temperature is above 40°C. The initial temperature required to achieve an internal temperature of 45°C at the time of semen collection depends on the environmental temperature. In most cases initial water temperature of 55° to 60°C will be adequate. The internal temperature should be checked periodically with an accurate thermometer if collection is delayed or very low temperatures are experienced in the collection area.

The collection funnel and tube must be maintained at near body temperature during collection. An insulated jacket which can be prewarmed should be attached to the end of the casing. A zippered opening is desirable so that the ejaculated sample can be examined. At the time of ejaculation the glans penis of most bulls coils sufficiently to cause the collection funnel and tube to be twirled. The insulated jacket thus protects the collection tube from possible breakage.

The AV must be lubricated before it is used to collect semen. The open end and one-third of the inner liner is lubricated with a water soluble surgical jelly using a sterile glass rod. The penis carries enough lubricant forward to lubricate the remainder of the liner. The jelly should be used sparingly since an excessive quantity will contaminate the semen, causing clumping of sperm.

**14-2.2a Collecting from the bull.** The AV method of collecting semen from the bull involves the use of properly restrained teaser-mount animals. Dummy mounts have been constructed but are seldom used. It is necessary for the bull to mount in order for him to serve the AV and ejaculate. The teaser-mount animal serves this purpose but in addition provides the sexual stimulus necessary to get the bull sexually prepared for ejaculation. Cows, bulls, or steers may be used as teaser-mounts.

When collecting from the right side of the bull, the AV should be balanced in the right hand with the palm up and held parallel to the expected path of the bull's penis. As the bull mounts and the penis is extended, the left hand is used to guide the penis to the side by grasping the sheath. The extended penis should not be touched by the hand because this results in either a retraction of the penis or ejaculation before the penis enters the AV (Figure 14-3).

Best ejaculation results are obtained when the penis is covered by the AV on the upward movement as soon as the bull mounts. It is helpful for the attendant to place his left shoulder against the side of the bull as he mounts so their movements are in unison (Figure 14-4). Accurate timing is essential for best results. When the bull thrusts for ejaculation, the AV should be allowed to move with the thrust and maintained as near in line with the penis as possible. The AV should be held on the penis until the bull begins to dismount. Sharp bends of the penis which may cause discomfort, or even injury, should be avoided. An experienced semen collector soon learns individual differences between bulls and will alter the collection procedure accordingly. Figure 14-5 shows an ejaculate of bull semen in the protective covering.

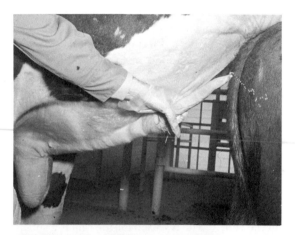

FIGURE 14-3.   Grasping the bull's penis rather than the sheath may cause premature ejaculation. (Courtesy of Select Sires, Inc., Plain City, Ohio.)

FIGURE 14-4.   Semen collection from the bull using the artificial vagina. (Courtesy of Select Sires, Inc., Plain City, Ohio.)

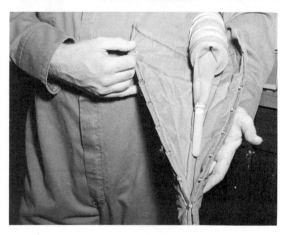

FIGURE 14-5.   An ejaculate of bull semen. Note protective covering for the funnel and collection tube which prevents cold shock and tube breakage. (Courtesy of Select Sires, Inc., Plain City, Ohio.)

Several precautions must be taken at the time of collection to insure high quality semen:

1. Cold shock is prevented by providing adequate protection for the collecting funnel and tube.
2. The collection tube must be protected from sunlight.
3. Contamination of semen by urine, water, or lubricating jelly must be avoided.
4. Microbial contamination is minimized by using a different sterile artificial vagina for each ejaculation.

A clean AV should be used even when a bull fails to ejaculate after the AV is placed over the penis. Second ejaculates collected with the same AV contain twice as many bacteria as first ejaculates. The number of bacteria also increases greatly with the number of insertions of the penis into the artificial vagina. Preputial hairs should be clipped as an aid to sanitation.

**14-2.2b Collecting from the ram.** The AV used for the ram is a smaller version of the one used for the bull. The recommended temperature and method for collection are the same as that for the bull. The collector must have quick reflexes since the ram mounts and serves rapidly. The AV can be attached to a dummy with a sheep skin stretched over it. The ram mounts the dummy with little training. The AV is the preferred method for collecting ram semen for use in AI.

**14-2.2c Collecting from the stallion.** The AV for the stallion must be larger than that for the other farm species. The outer casing is usually made of leather and is 75 to 80 cm in length (Figure 14-6). It must have a handle or strap to help the collector support its weight and to cope with the vigorous thrusting of the stallion. The inner liner and some version of the collection funnel and tube are also used. The AV is only partially filled with water to provide an internal temperature of 45°C. A valve designed to relieve pressure as the penis enters is essential. Sufficient pressure and friction must be maintained to stimulate ejaculation. The penis should be washed with warm soapy water and rinsed prior to ejaculation to remove debris from its surface.

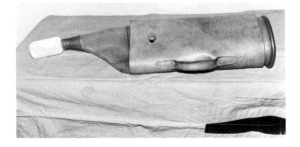

FIGURE 14-6. Artificial vagina used to collect semen from the stallion.

Steps to follow in collecting semen from a stallion include the following:

1.  Tease him to get extension of his penis.
2.  Wash his penis with water and a mild soap followed by complete rinsing.
3.  Have stallion mount a dummy or the estrous mare from an angle.
4.  Divert penis into the artificial vagina with a hand.
5.  If he is mounting a mare brace the artificial vagina against the thigh of the mare and hold it parallel to the direction of his thrust.

The outer end of the AV should be lowered sufficiently to allow semen to flow into the collection bottle. Ejaculation is complete in about 12 to 25 seconds.

**14-2.2d Collection from the boar.**   The collection of semen from the boar requires an entirely different procedure. In natural service the boar ejaculates when the corkscrew shaped glans penis is firmly engaged in the sow's cervix. Pressure of this engagement stimulates ejaculation. If an artificial vagina is used to collect semen from the boar, it must be designed so that the glans penis receives adequate pressure. Such artificial vaginas have been successfully used. The preferred method of collecting semen from the boar involves grasping the glans penis with a gloved hand as it emerges from the sheath (Figure 14-7). Continuous pressure is applied to stimulate ejaculation which may last from 10 to 20 minutes. Ejaculation is interrupted if pressure is not maintained.

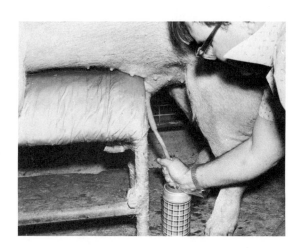

FIGURE 14-7.   Semen collection from the boar. The gloved hand replaces the artificial vagina.

**14-2.2e Cleaning, sterilizing and storing the AV.**   Washing and sterilizing the artificial vagina are extremely important. It should be completely disassembled, rinsed with tap water, and washed immediately. If washing cannot be accomplished immediately the parts should soak in lukewarm water until washing can be completed. The casings should be brushed with a stiff brush and water containing a detergent. They are rinsed with tap water and allowed to air dry. Inner liners and funnels should be turned inside out and brushed thoroughly with the detergent water. They are rinsed with tap water and then with distilled water prior to soaking for 5 minutes in 70% isopropyl or ethyl alcohol. Some prefer to boil the liners and funnels in distilled water for 15 minutes before rinsing with 70% alcohol. In the latter case, the alcohol would serve primarily as a drying agent. After the alcohol rinse, the parts should be hung in a dust-free cabinet to dry. If used daily they may remain in the cabinet; otherwise each piece can be individually wrapped in clean paper towels for storage.

**14-2.3**
**Electroejaculation**
 This method of collecting semen is preferred to the AV method under certain conditions. It is used for dairy bulls that have become crippled, have low sexual activity due to age, or for other reasons are unable to serve the artificial vagina. It is extensively used with beef bulls, especially to check semen quality on bulls to be used in natural mating prior to the breeding season. Semen should not be collected and used from males that have not demonstrated normal sexual behavior or ability to ejaculate, as the cause may be genetic and transmitted to the offspring.

The equipment used for electroejaculating males consists of a bipolar electrode and a variable source of alternating current. The voltage ranges from 0 to 30 with a low amperage (0.5 to 1.0). The electrode may have either alternating positive and negative rings spaced 4 cm apart or four conductors, two positive and two negative, running longitudinally along the electrode. One model has three conductors, all on one side. (See Figure 14-8.) The electrode is placed in the rectum immediately above the accessory glands so that the nerves of the reproductive system are stimulated.

FIGURE 14-8. Electroejaculator used to collect semen from the bull and the ram. Small probe in the center is used on the ram.

**14-2.3a Electroejaculation of the Bull.**  Remove excess fecal material from the rectum, insert the lubricated electrode, and position it immediately over the accessory glands. Stimulation is begun with low voltage which is gradually increased, a few volts at a time, alternated with 4-second rest periods in which the voltage is returned to 0. This rhythmic pattern is continued until the bull is stimulated to ejaculate. The secretion of bulbo-urethral fluid and penile erection should take place at the lower voltage, with ejaculation occurring at the higher voltage levels. Increasing the voltage too rapidly can result in ejaculation without erection and the semen will be contaminated by the prepuce. Most bulls are stimulated to ejaculate in a period ranging from 2 to 5 minutes.

The nerves of the rear legs are also stimulated by the electrical current. This results in stiffening of the legs and in some cases the bull goes down. It is essential that the bull be restrained in a chute, with good footing provided. In some cases it is desirable to provide some support under the ribcage to help support the bull's weight.

Ejaculates obtained by the electroejaculation method are usually larger in volume but lower in concentration than those obtained with the artificial vagina. The total number of sperm and the fertilizing capacity are about equal for the two methods.

**14-2.3b Electroejaculation of the ram.**   The ram responds to a lower voltage and also responds faster than compared to the bull. Ejaculation can be accomplished with the ram lying on a table or in a standing position. Usually three levels of voltage are required, with a peak of 8 volts. Some rams fail to respond, while others yield low quality semen. Since the urethral orifice is continuous with the filiform appendage, it is necessary to direct the latter into the collection tube before ejaculation.

**14-2.3c Electroejaculation of the boar.**   The level of voltage required for ejaculation is greater in the boar than in either the bull or the ram. This is probably due to the insulating effect of the body fat. The high voltage results in considerable discomfort, making some type of anesthesia desirable. Sperm motility, fertility, and concentration of electroejaculated boar sperm are good but the volume is usually low. The method is not satisfactory for routine collection of semen from boars.

**Suggested reading**

Almquist, J. O. 1973. Effects of sexual preparation on sperm output, semen characteristics and sexual activity of beef bulls with a comparison to dairy bulls. *J. Anim. Sci.*, 36:331.

Almquist, J. O. 1978. Bull semen collection procedures to maximize output of semen. *Proc. 7th Tech. Conf. Artif. Insem. Reprod.*, p. 33. NAAB, Columbia, Mo.

Ball, L. 1978. Semen collection by electro-ejaculation and massage of the pelvic organs. *Proc. 7th Tech. Conf. Artif. Insem. Reprod.*, p. 57. NAAB, Columbia, Mo.

Branton, C., G. D'Arensbourg and J. E. Johnston. 1952. Semen production, fructose content of semen and fertility of dairy bulls related to sexual excitement. *J. Dairy Sci.*, 35:801.

Collins, W. J., R. W. Bratton and C. R. Henderson. 1951. The relationship of semen production to sexual excitement of dairy cattle. *J. Dairy Sci.*, 34:224.

Foster, J., J. O. Almquist and R. C. Martig. 1970. Reproductive capacity of beef bulls. IV. Changes in sexual behavior and semen characteristics among successive ejaculates. *J. Anim. Sci.*, 30:245.

Herman, H. A. and F. W. Madden. 1972. *The Artificial Insemination of Dairy and Beef Cattle.* (4th ed.) Lucas Brothers Publishers. Columbia, Mo.

Pickett, B. W. 1968. Collection and evaluation of stallion semen. *Proc. 2nd Tech. Conf. Artif. Insem. Reprod.*, p. 80. NAAB, Columbia, Mo.

Salisbury, G. W., N. L. VanDemark and J. R. Lodge. 1978. *Physiology of Reproduction and Artificial Insemination of Cattle.* (2nd ed.) W. H. Freeman and Co.

# Chapter

# 15

# Semen evaluation

The fertility level of an ejaculate of semen is the ultimate in its evaluation. This can be accomplished only by inseminating females and waiting long enough to determine whether or not they return to estrus or until a pregnancy diagnosis can be made. Most AI organizations use a *nonreturn rate* as a measure of fertility for their bulls. Nonreturn rate is the percentage of females not reported as having had a second insemination within a given period of time. The earliest such information can be made available is about 35 days, and more reliable data are not available until 90 days after the semen is first used in the field. Even with frozen semen, a high percentage of an ejaculate may be used before any information is available on its actual conception rate.

It is imperative, therefore, that laboratory technicians learn as much about an ejaculate of semen as possible within the framework of economic feasibility. This chapter is devoted to the several laboratory tests which, individually or in combination, give some indication of semen quality. The student should keep in mind that these tests are not actual measures of fertility and that an individual ejaculate that scores high on the laboratory tests may actually have a lower fertility rate than another sample that scores lower. However, the average conception rate of all ejaculates that score high on the laboratory tests will be higher than the average fertility rate for those that score low. The procedures for conducting each individual test will be described and the value and limitations of each will be discussed.

**15-1
Gross
examination**

Each ejaculate of semen is examined immediately for certain characteristics. The volume is determined not only for use in processing but also to establish a pattern for the individual bull. Deviations from this pattern, particularly downward trends in volume, indicate a problem. The problem may be due to health factors or it could be an indication that the collection procedures for that particular bull need revision.

The gross appearance of a normal ejaculate of bull semen is creamy white in color. When observed closely swirls of movement can be seen. The layer along the glass tube may have a granular appearance due to sperm movement. Samples with low concentrations will appear watery or less opaque. Ram semen, while lower in volume and higher in concentration, has a similar appearance to bull semen. Stallion and boar semen, being much less concentrated, will not have the dense appearance of bull and ram semen.

After standing for a few minutes the bottom of the collection tube or bottle should be examined for dirt or other debris. Presence of such contaminants would indicate that the male had not been properly cleaned prior to collection or that careless collection techniques exist. (See Figure 15-1.)

FIGURE 15-1. Freshly collected ejaculates of (L to R) ram, bull, stallion, and boar semen. Note differences in volume and density.

**15-2
Progressive
motility**

Motility of a sample of semen is expressed as the percentage of cells that are motile under their own power. A progressively motile sperm is one that is moving or progressing from one point to another in a more or less straight line. Most ejaculates will show other types of motility. These include both circular and reverse movement due to a tail abnormality and vibrating or rocking movement often associated with aging. Progressive motility is the most important individual quality test because fertility is highly correlated with number of motile sperm inseminated. The percentage motility of semen ejaculates can range from 0 to 80%. The usual motility determination is a subjective measurement based on the judgment of the individual making the determination. Because of this, most laboratory personnel will use multiples of ten in reporting progressive motility (that is, 40, 50, 60, or 70%) feeling that this is as accurate as one can make the determination. Based on large numbers of ejaculates there is little difference in the fertility level of ejaculates with initial motilities of 50 to 80% if the desired number of motile sperm are present at the time of insemination. Any difference that may exist is probably masked by the usual practice of adding additional sperm per breeding unit to bring the number of motile sperm to optimum. Samples with less than 40% initial motility are not suitable for use unless the ejaculate is from an exceptionally superior bull from which one would be willing to accept a low conception rate. It appears that the motile sperm present in these low motility samples have had their fertilizing capacity affected by the factor or factors that caused the low motility.

15-2.1
Preparation of
slide

Some precautionary measures should be taken prior to the preparation of the slide. If the sample has been standing for several minutes it should be mixed by gently inverting the tube two or three times. Undiluted semen, particularly from the bull and the ram, is too concentrated for accurate motility determinations. Small subsamples must be diluted with an *isotonic* solution (same free ion concentration as the semen) so that individual sperm can be observed. This requires a dilution rate of about 1 to 100 for bull and ram semen and 1 to 10 for boar and stallion semen.

Motility examinations must be made with the aid of the microscope and a suitable preparation on a microscope slide. A clean dry slide should be placed in a stage incubator or on a heated microscope stage and allowed to warm to a temperature of 38°C. A small droplet of diluted semen is placed on the slide using a glass stirring rod or small pipette. It is important that uniform droplets be used in each preparation so that accurate determinations can be made. The droplet should be covered with a glass cover slip to spread to a uniform thickness and to prevent drying. The slide is now ready for microscopic examination.

15-2.2
Use of
microscope

The prepared slide should remain in the stage incubator during examination to prevent changes in temperature (Figure 15-2). A magnification of approximately 400× is preferred. Several fields should be examined and an estimate made of the percentage of sperm that are progressively motile to the nearest 10%. For samples that are above 50% motile, it is easier to make the estimate from the percentage of sperm not progressively motile. Equipment is available that will display the microscopic image on a glass screen making the determination easier and more accurate.

Motility determinations are made on diluted samples both before and after freezing in the same manner as described for fresh semen, except that it is not necessary to further dilute the samples before preparing the slide. It is important to remember that the diluted sample contains either egg yolk or milk and in the case of frozen semen, glycerol. The viscosity contributed by these materials slows motility but does not alter the percentage of progressively motile sperm. Whole milk diluters and some others such as egg yolk with phosphate buffer make the individual sperm cells difficult to see. Motility percentage can be determined in these materials but it is more difficult and less accurate.

FIGURE 15-2. Stage incubator or heated microscope stage keeps semen smear warm while motility determination is made.

Circular and backwards motility are indications of either cold shock or contamination with water. *Hypotonic* (lower free ion concentration than semen) diluters due to error in preparation can also cause this problem. When this abnormal motility pattern is observed, check these three factors in the order listed. See Chapters 14 and 16 for additional information.

## 15-3
## Concentration
## of sperm cells

Sperm cell concentrations are expressed as number of cells per cc and must be known on each ejaculate in order to make the maximum number of breeding units containing a given number of motile sperm per unit. A permanent record should be kept of the concentration of each ejaculate from each bull. A downward trend in concentration for a particular bull might indicate a serious problem. The problem could be related to insufficient sexual stimulation prior to collection, a collection schedule that is too strenuous, or an illness that occurred or began several weeks earlier. There is little relationship between concentration and fertility when the concentration is within the normal range. Again, some masking may occur due to compensating dilution rates. Ejaculates from a bull containing less than 500 million cells per cc should be used with caution because lowered fertility is likely. There is less information available on the other species, but caution is advised when the sperm cell concentration drops below 50% of normal.

### 15-3.1
### Direct cell count
### (hemocytometer)

The hemocytometer was designed for counting blood cells. It consists of a specially designed slide that contains two counting chambers and two dilution pipettes. The counting chambers are 0.1 mm in depth and have a ruled area on the bottom of the chamber that is 1.0 mm square. This square is subdivided into 25 smaller squares (Figure 15-3). With a known depth and area, the number of sperm can be determined for a specific volume. One of the 2 pipettes is designated for counting red blood cells and is designed for dilution rates of 1:100 and 1:200. A dilution rate of 1:200 is normally used for bull and ram semen while a dilution rate of 1:100 is preferable for gel-free

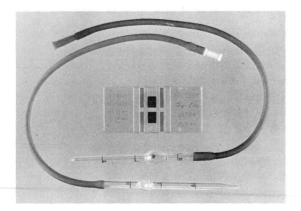

FIGURE 15-3. Standard hemocytometer slide and dilution pipettes. Large pipette is used for bull and ram semen, while small one is used for boar and stallion semen.

boar and stallion semen. The smaller pipette may be used on boar and stallion semen of low concentration. The diluent used must kill the sperm so that counting can be accomplished. The authors have found that a 2% aqueous solution of eosin accomplishes this and has the additional advantage of staining the sperm heads so that they are easier to count. Semen is drawn into the capillary tube of the dilution pipette to the desired mark. The pipette tip is carefully wiped to remove semen clinging to the outside without drawing any from the capillary. The semen is pulled into the bulb of the pipette and the pipette is filled with the diluent. The contents of the pipette are mixed either with a mechanical shaker or by holding the ends of the pipette between the thumb and the index finger and shaking it vigorously in 100 back-and-forth 12-inch movements. Sufficient diluted semen is blown from the dilution pipette to insure that the diluent containing no sperm has been flushed from the capillary.

The cover slip is placed over the counting chambers and a small amount of semen is placed at the edge of the cover slip. Capillary action will draw the semen under the cover slip. Both chambers are thus filled being careful not to overfill. Overfilling results in semen running off the edge of the counting chamber setting up currents which make counting difficult and inaccurate. The slide is placed under the microscope and the scribed area of a counting chamber is located using the low power objective (100×). The 1.0 mm square area is examined for uniform distribution of sperm. The high power objective (400×) is used to count the number of sperm within the desired number of squares. With bull and ram semen the corner and center squares are counted. With boar and stallion semen it might be necessary to count the entire twenty-five squares. Both chambers can be counted for greater accuracy. When twenty-five squares are counted the volume represented is 0.1 mm³. If only five squares are counted, the number must be multiplied by 5 to obtain the number of sperm in 0.1 mm³. The number of sperm/cc can be found by using the following formula:

$$\text{No. sperm/cc} = \text{No. sperm in } 0.1 \text{ mm}^3 \times 10 \times \text{dilution rate} \times 1000$$

The hemocytometer is a time consuming and tedious method of determining sperm concentration. It is also subject to considerable error for the inexperienced or careless technician. However, the conscientious individual with some experience can provide accurate information. The inexperienced person should make two dilutions of a sample and count two chambers from each dilution as a check on his technique. This should be continued until the two dilutions consistently agree within 5% of each other.

**15-3.2
Photoelectric
colorimeter**

This is the method used by commercial semen producing operations. A determination can be made in approximately 1 minute. The procedure is based on light absorption by the sperm in the semen sample. The equipment consists of a light source which passes through a series of lenses and filters and then through a diluted sample of semen. The light which passes through the semen is measured with a galvonometer (see Figure 15-4). There are several makes and models of instruments suitable for use but each must be calibrated using samples of known concentration as determined by hemocytometer, so that the galvonometer readings can be converted to concentration figures.

A concentration determination is made by pipetting the required volume of 2.9% sodium citrate dihydrate solution into a vial especially designed for the instrument. To this is added the required amount of the ejaculate. The tube is mixed by inverting two or three times before inserting into the instrument. The instrument is turned on and allowed to warm up before use. A reading is taken from the galvonometer and referred to the calibration table for the concentration figure. Since the sodium citrate solution used to dilute the semen for this determination is a part of most semen diluters this subsample may be added to the diluted ejaculate provided it has been handled properly and not allowed to become cold shocked.

FIGURE 15-4. One model of a photoelectric colorimeter used to determine semen concentration.

**15-3.3
Electronic
particle counter**

This instrument can be used to accurately determine the number of sperm cells in an ejaculate (Figure 15-5). It is more accurate than either the hemocytometer or the photoelectric colorimeter. The instrument can be adjusted for particle size so that only the sperm cells in a sample will be counted. A diluted sample of semen is passed through a capillary so that only one cell at a time can pass between two electrodes. The sperm head causes an abrupt increase in resistance which is registered on a counter. The greatest disadvantage of this instrument is its cost. It is not routinely used in semen processing.

FIGURE 15-5. Electronic particle counters may be used to determine sperm cell concentration.

**15-4
Sperm cell
morphology**

Sperm cell morphology was covered in Chapter 12 and the student is referred to Figure 12-2 for the various sperm abnormalities. The abnormal sperm count is not conducted on every ejaculate of semen because of the time required. A count should be made monthly so that a trend in percentage of abnormal sperm can be charted on each bull. Sudden increases in percentage of abnormals would indicate a problem which should be followed carefully. Semen ejaculates may show as few as 5% abnormal sperm while others may approach 100%. Fertility is usually not affected until the level of abnormal sperm exceeds 20 to 25%. Abnormal sperm do not show progressive motility. Therefore, as the percentage of abnormal sperm increases the progressive motility percentage would decrease. Many abnormal sperm can be detected while examining samples for motility. This can be used as an indication that further morphological study is needed.

**15-4.1
Preparation of
slide**

It is essential that good technique be used in preparing the slides for morphological examination. Poor technique can result in damage to the tails, which could be interpreted as naturally occurring abnormalities. This can be quite serious since a high percentage of the observed abnormalities involve the tail. If a freshly collected sample is to be examined, a small drop of 2.9% sodium citrate solution is placed on a warm microscope slide. A 3-mm glass stirring rod is touched to the surface of the semen sample, and the sperm that adhere to the rod are transferred to the citrate solution on the slide. The semen is carefully mixed with the citrate solution and another warm slide is placed flat over the first, spreading the mixture evenly between them. The two slides are separated by pulling the ends in opposite direction with a smooth motion (Figure 15-6). Care must be taken to have sufficient fluid between the two slides to prevent pressure on the sperm. One of the slides should be placed flat on a warm surface and allowed to dry.

Slide preparations from diluted semen are made in the same manner, except the dilution rate of the semen must be considered. Samples that are diluted for maximum utilization usually do not need further dilution with sodium citrate solution. The objective of dilution is to disperse the sperm sufficiently so that individual sperm may be observed. If the sperm are too concentrated on the first slide, a second one can be made with the appropriate dilution.

FIGURE 15-6. Slide preparation for sperm morphology determination. Semen and staining mixture is spread between two microscope slides. The slides are gently pulled apart and dried.

### 15-4.2
### Staining slides

A large number of staining materials and techniques have been used. The principal objective of any staining technique is to make the cells easily observed so that abnormalities can be discerned. The easiest and most widely used techniques of staining incorporates the stains in the preparation of the slide. Eosin, along with a background stain such as nigrosin, opal blue, or a fast green, works well (see section 15-5.1). Nigrosin is superior to opal blue and fast green for diluted samples containing egg yolk. Eosin does not stain the living sperm heads but the background stain causes the unstained sperm heads to be clearly visible.

Other staining and counter staining procedures have been used. Several variations using carbol fuchsin as a stain, with methylene blue or analine gentian violet as a counter stain, have been used successfully.

**15-4.3
Examination
and counting**

Several fields on each slide should be examined to determine whether a satisfactory slide has been prepared. The sperm should be dispersed so that the individual heads and tails can be observed clearly. A high percentage of bent or broken tails could indicate poor technique in slide preparation. When most of the abnormal tail bends are oriented in the same direction one should suspect that abnormal pressure due to insufficient fluid between the two slides occurred during preparation. Should either of these problems be observed another slide should be made with greater care. The slide should be examined with a magnification of approximately 400×.

One hundred sperm should be observed at random per slide and classified. Statistical studies have shown that little precision is gained by counting more than 100 sperm per slide. If more precision is desired, more than one slide should be made and counted on the same sample. (See Section 12-1 for the classification of morphologically abnormal sperm.)

**15-5
Differential
staining of live
and dead
sperm**

Eosin is referred to as a differential stain in that it cannot pass through living cell membranes but can pass through nonliving cell membranes. A background stain such as nigrosin, opal blue, or fast green helps make the unstained sperm heads visible. The percentage of live sperm in a sample of semen has been used as a verification of motility determinations. The student should keep in mind that the percentage of live sperm will always be somewhat higher than the percent motility.

**15-5.1
Staining
techniques**

Several staining mixtures have given good results. All of them contain about 1% eosin plus one of the background stains. Preferred background stains are 2% fast green, 4% analine blue, or 5% nigrosin. Both eosin and the background stain are dissolved in 2.9% sodium citrate dihydrate buffer. The slides are prepared in the manner described in Section 15-4.1 with two exceptions: (1) the staining mixture is substituted for the sodium citrate buffer used to dilute the semen; (2) the slides must be dried quickly on a heated plate (55 to 60°C) with a small electric fan directed across the plate (Figure 15-7). When slides are allowed to dry slowly some of the sperm may die and be stained before the drying process is completed, thus giving a false indication of the percentage of live sperm.

**15-5.2
Counting slides**

Several fields at random should be examined and counted. Between 300 and 500 total cells are usually counted in order to give accurate results. Partially stained sperm should be included with the totally stained sperm representing the number of dead sperm in the sample. The unstained sperm represents those that were alive in the sample.

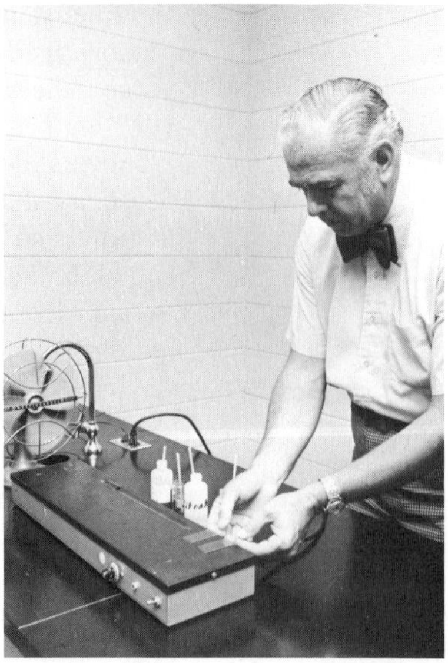

FIGURE 15-7. Slides for live-dead determination must be dried quickly using heated plate and electric fan.

**15-6
Speed of
sperm**

The rate of motility is defined as the speed at which sperm travel. Electronic equipment is available which can measure sperm speeds objectively, but the equipment is expensive and is not used in a routine semen producing operation. The rate of motility is frequently assessed subjectively on a scale of 1 to 5. This can be done about as accurately as the motility percentage but has little value in evaluating semen quality.

Individual sperm have been clocked at speeds of 21 mm per minute and in samples of semen with excellent motility several sperm have shown speeds averaging 15 mm per minute. When several sperm are checked from each of a large number of ejaculates, an average speed of 6 mm per minute can be expected.

**15-7
Evaluating
frozen semen**

The previously described quality tests must be used to determine whether an ejaculate of semen is good enough to process. A second series of evaluations must be made following freezing. The freeze-thaw survival rates of semen from different bulls and different ejaculates from the same bulls vary greatly. Since conception rate in AI depends greatly on an adequate number of motile sperm at the time of insemination, it is imperative that an accurate post-thaw evaluation be made. Each ejaculate of frozen semen must be evaluated to determine whether it will provide that optimum number of motile sperm. Between 5 and 15% of the ejaculates meeting the standards for processing are discarded after freezing. Some of the variation in the percentage of ejaculates discarded seems to be related to season with the highest rate in the summer.

**15-7.1
Photographic
method of
determining
progressive
motility**

Time-exposure photography as an objective method of determining motility has been reported. The procedure employs a dark-field microscope equipped with a camera. Semen should be diluted to 10 million sperm per cc and placed in a Petroff-Hausser counting chamber. This chamber has a depth of only 20 microns compared to 100 microns of the hemocytometer chamber and conforms better to the depth of focus of the microscope.

The filled chamber is warmed for 30 seconds on a 38°C warming plate before it is placed on the heated stage of the dark-field microscope. Six 2-second exposures are made at the periphery of the ruled area at the 1, 3, 5, 7, 9, and 11 o'clock positions. The desired magnification to the film is 75×. The developed film is projected on a white wall for further magnification and counting (Figure 15-8). Motile cell counts are made by counting the number of tracks on the film. Nonmotile cells must also be counted to determine percentage motility. The procedure gives highly repeatable results.

Motility can be determined photographically at any stage of processing. However, the procedure has been used primarily in determining post-freezing motility. This is probably due to the time and expense involved. An efficient laboratory technician can make thirty to thirty-five determinations in an 8-hour day.

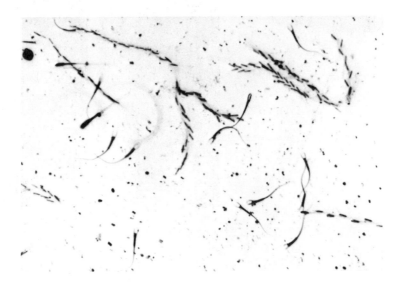

FIGURE 15-8. Time-exposed photomicrograph used to objectively measure pro-gressive motility. Streaks represent motile sperm. (Courtesy of American Breeders Service, DeForest, Wis.)

**15-7.2
Changes in the
acrosome**

Researchers have reported that aging of or injury to sperm causes acrosomal cap deterioration. Changes are first noted at the apical ridge but progresses to the point where the entire acrosome is gone. These changes were first noted with the electron microscope but can be seen with differential interference contrast and phase contrast optics. Differential interference contrast seems to be preferred. (See Figure 15-9.)

The freezing and thawing of semen inflicts injury to some sperm. The important point is to know accurately the percentage of injured sperm. Good quality dairy bull semen will have about 90% intact acrosomes prior to freezing. After freezing the intact acrosomes drop to 60 to 65%. Most beef bulls will have 50 to 55% intact acrosomes in post-thaw samples. Studies have shown higher correlations between intact acrosomes and nonreturns than between post-thaw motility and nonreturns.

To determine damaged acrosomes, thawed semen is incubated in a 37°C water bath for 2 to 4 hours. A drop of incubated semen is placed on a warm slide and covered with a cover slip. Motility and acrosome integrity are evaluated by counting 100 sperm using the differential interference contrast microscope at 800×. If the first sample is substandard, a second smear should be made and examined. If it is also substandard, another sample may be thawed and examined the next day before the ejaculate is discarded.

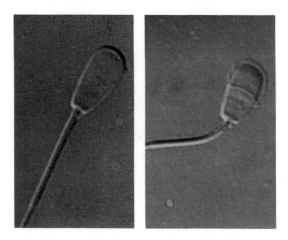

FIGURE 15-9. Photomicrograph of post-thawed spermatozoa; intact acrosome (left) and damaged acrosome (right). (Courtesy of Select Sires, Inc., Plain City, Ohio.)

There is no minimum percentage of intact acrosomes to use in discarding ejaculates. A profile must be developed for breeds and then for individual bulls. When the percentage of intact acrosomes falls significantly below this profile for an ejaculate it should be discarded.

The use of intact acrosomes procedure has been adopted much more widely than the photographic method of determining motility. Many commercial semen producing businesses are examining acrosomes on about 50% of their ejaculates and some are examining all ejaculates after freezing.

**15-8
Other tests**

Several other tests have been suggested and used in evaluating semen samples. Livability in storage was used extensively when semen was utilized in the liquid form. This involved daily motility determinations on the semen stored at 5°C. The freezing of semen has largely eliminated this procedure.

The resistance of sperm to cold shock has been used as a measure of quality. Several methods of measuring metabolic activity such as oxygen uptake, fructolysis, methylene blue reduction time, resazurin reduction time, and pH change all provide some information.

**Suggested
reading**

Blom, E. 1950. A one-minute live-dead sperm stain by means of eosin-nigrosin. *Fert. and Ster.*, 1:176.

Elliott, F. I. 1978. Semen evaluation. *Physiology of Reproduction and Artificial Insemination of Cattle*. (2nd ed.) ed. G. W. Salisbury, N. L. VanDemark and J. R. Lodge. W. H. Freeman Co.

Foote, R. H. 1972. How to measure sperm cell concentration by turbidity. *Proc. 4th Tech. Conf. Artif. Insem. Reprod.*, p. 57. NAAB, Columbia, Mo.

Marshall, C. 1978. Use of differential interference contrast optics as a quality control and research tool. *Proc. 7th Tech. Conf. Artif. Insem. Reprod.*, p. 62. NAAB, Columbia, Mo.

Mitchell, J., R. D. Hanson and N. Fleming. 1978. Utilizing differential interference contrast microscopy for evaluating abnormal spermatozoa. *Proc. 7th Tech. Conf. Artif. Insem. Reprod.*, p. 64. NAAB, Columbia, Mo.

Saake, R. G. 1972. Semen quality tests and their relationship to fertility. *Proc. 4th Tech. Conf. Artif. Insem. Reprod.*, p. 22. NAAB, Columbia, Mo.

Swanson, E. W. and H. J. Bearden. 1951. An eosin-nigrosin stain for differentiating live and dead bovine spermatozoa. *J. Anim. Sci.*, 10:981.

Van Dellen, G. and F. I. Elliott. 1978. Procedure for time exposure darkfield photomicrography to measure percentage progressively motile spermatozoa. *Proc. 7th Tech. Conf. Artif. Insem. Reprod.*, p. 55. NAAB, Columbia, Mo.

# Chapter

# 16 Semen processing, storage, thawing, and handling

**16-1**
**Importance**
**and properties**
**of semen**
**diluters**

The success of artificial insemination (AI), particularly in cattle and sheep, depended greatly on the development of satisfactory semen diluters. The AI industry has adopted the term extender to replace diluter. When AI was first being adopted some felt that the use of the term "diluted semen" created a stigma, so a switch to "extended semen" was made. The authors prefer the original term, diluter. A 6-cc ejaculate of semen from the bull may contain sufficient number of motile sperm to inseminate 200 to 300 cows, but it would be virtually impossible to divide 6 cc into that many units. The AI pioneers also found that sperm in undiluted semen lived for only short periods of time and that cooling undiluted semen very slowly to 5°C caused the death of many sperm. It became obvious that a satisfactory semen diluter would have to do more than increase the volume of an ejaculate of semen. The diluter would have to protect the sperm during cooling and extend the life of the sperm.

The following properties of a good semen diluter have been delineated along with examples of materials that satisfy these properties.

1. A diluter must be isotonic with semen (have the same free ion concentration)—2.9% sodium citrate dihydrate.

2. Buffering capacity must be provided (prevent pH change by neutralizing acid produced by sperm metabolism)—isotonic sodium citrate solution.

3. Diluters must protect the sperm from cold shock injury during the cooling from body temperature to 5°C—lecithin and lipoproteins from egg yolk or milk.

4. Nutrients must be provided for sperm metabolism—egg yolk, milk and some simple sugars.

5. Microbial contaminants must be controlled—antibiotics such as penicillin and streptomycin.

6. Sperm must be protected from injury during freezing and thawing—glycerol.

7. The diluter must preserve the life of the sperm with a minimum drop in fertility—combination of known and unknown factors.

## 16-2
## Buffer
## solutions used
## in semen
## diluters

The buffer solution as described in the preceding paragraph serves a dual role. By neutralizing the acid produced by the metabolic activity of the sperm, minute changes in pH are prevented. The proper concentration of the buffer salt provides an isotonic environment for the sperm. In addition the salt used must not be toxic to the sperm at the level required for isotonicity. Of the many compounds and combinations of compounds available that satisfy at least one of the three criteria listed, only three buffer solutions have been found to be satisfactory for use in semen diluters.

## 16-2.1
## Phosphate
## buffer solution

The phosphate buffer was a component of the first satisfactory semen diluter reported in 1939. It was composed of 2.0 gm of $Na_2HPO_4 \cdot 12\ H_2O$ and 0.2 gm of $KH_2PO_4$ in sufficient distilled water to make 100 cc of solution. While just as satisfactory as other buffer solutions, phosphate buffer has not been as popular because it produces an opaque mixture when added to egg yolk, resulting in poor sperm visibility.

## 16-2.2
## Citrate buffer
## solution

The suitability of sodium citrate dihydrate solution as a suitable buffer for semen was discovered in 1941. It is composed of 2.9 gm of sodium citrate dihydrate in sufficient distilled water to make 100 cc of solution. The sodium citrate buffer soon replaced the phosphate buffer in preparing semen diluters. When mixed with egg yolk it leaves the mixture sufficiently transparent to give good visibility of the individual sperm.

## 16-2.3
## Tris buffer
## solution

Tris (hydroxymethyl) aminomethane has been extensively studied as a buffered medium for bull and boar sperm. The tris buffer seems to have value in prolonging sperm life at ambient temperature, 5°C and −196°C. Various molarities and pH levels have been tested. A 0.2 m concentration and a pH of 6.5 plus 1% fructose gives best results. The buffer is prepared by adding 3.028 g tris, 1.678 g citric acid and 1.0 g fructose to sufficient distilled water to make 100 cc of buffer.

## 16-3
## Antimicrobial
## agents for
## semen diluters

Attention was called to the problem of microbial contaminants in ejaculated bull semen as early as 1941. The number of organisms per cc can range from no detectable organisms to several million. The number of organisms can be reduced by proper cleaning of the sheath and under line of the male prior to collection. All equipment used in semen collection, processing, and storing should be sterile and not contributors of other contaminants.

A wide variety of organisms has been isolated from semen. Many of these organisms are not pathogenic but do compete with the sperm for nutrients and do produce metabolic by-products that have an adverse effect on livability of the sperm. Fortunately, it has been the longtime goal of the commercial AI organizations to eliminate from their bulls the diseases that can be transmitted through the semen. (These specific diseases will be discussed in Chapter 25.) The possibility does exist that undiscovered pathogens may be ejaculated in the semen and unknown sources of extraneous contaminants may occur.

The beneficial effects of antibiotics added to semen diluters was discovered in 1946. Much of the increase in conception rate in those early years was the result of controlling the venereal disease vibriosis. Table 16-1 shows the effect of antibacterial agents on a group of low and a group of high fertility bulls. Even though the nonreturn rate was increased for the high fertility bulls, this increase was not nearly as dramatic as for the low fertility bulls. It was later determined that the primary difference in nonreturn rates for the two groups of bulls prior to the addition of antibacterial agents was due largely to *Vibrio fetus* infection in the low fertility bulls. In the absence of specific infectious disease organisms, antibiotics are beneficial by reducing competition from other bacteria.

A vast array of antibiotics and fungicidal agents have been used experimentally. Many of the antibiotics and particularly some of the fungicidal agents are extremely toxic to sperm. At levels compatible with sperm others are not very effective in controlling microbial contaminants.

The current recommendation for controlling microbial contamination in bull semen is as follows: sulfanilamide 0.3% in final diluter (liquid semen only), penicillin 1000 IU per cc of diluter, and streptomycin or dihydrostreptomycin 1000 $\mu$g per cc of diluter. Polymyxin B is added at the rate of 1000 $\mu$g per cc of diluter by some organizations but not by others. Researchers do not agree on the effectiveness of streptomycin without polymyxin B in controlling specific contaminants such as *Vibrio fetus* organisms.

TABLE 16-1. Fertility with antibacterial agents in 50% yolk-citrate: high and low fertility bulls

| | Increase in % 60-90 Day N R | |
| --- | --- | --- |
| | Low bulls | High bulls |
| No Antibiotic | 58% base | 65% base |
| Sulfanilamide | +3 | +1 |
| Penicillin | +10 | +6 |
| Streptomycin | +11 | +4 |
| Polymyxin | +3 | +2 |
| All Combined | +15 | +3 |

From Foote and Bratton, J. Dairy Sci. 33:544, 1950.

The addition of glycerol to diluted semen containing *Vibrio fetus* organisms has protected the organisms from the antibiotics. During the period when frozen semen was being adopted, many bulls were infected with *Vibrio fetus*. Research showed that waiting 6 hours after the addition of diluter containing 1000 $\mu$g of streptomycin per cc to semen from infected bulls before adding glycerol allowed the antibiotic to kill the Vibrio organisms. This procedure delayed the processing of the semen so other procedures were sought. A procedure of adding penicillin (1000 IU/cc), streptomycin (1000 $\mu$g/cc), and polymyxin B (1000 $\mu$g/cc) to freshly collected undiluted semen followed by 30 minutes of incubation at 30°C has proven to be effective in killing *Vibrio fetus* organisms experimentally introduced into the semen. The 30-minute incubation period prior to cooling has been shown to lower post-thaw motility but did not significantly lower nonreturn rate (NR). Most semen processors have not adopted the procedure probably because of the reduced post-thaw motility. They are relying on the cooling period to provide sufficient time for the streptomycin to kill the Vibrio organisms prior to the addition of the glycerol diluter. This lends support for the longer cooling time to be described later in this chapter.

## 16-4
## Effective semen diluters for bull semen

### 16-4.1
### The yolk-phosphate

The yolk phosphate diluter was prepared by mixing equal parts of the phosphate buffer solution and fresh egg yolk. The reaction of the phosphate ions on the fat globules of the egg yolk resulted in an opaque mixture making it impossible to observe individual sperm in the mixture. Even though this diluter maintains good motility and fertility of bull sperm, it is not often used.

### 16-4.2
### The yolk-citrate

The yolk-citrate diluter is prepared by adding fresh egg yolk to the citrate buffer solution. Prior to the adoption of frozen semen, the ratio of yolk to buffer solution was 1:1. Most of the reported research showed a slight decrease in nonreturn rate with lower percentages of egg yolk. When the semen is to be frozen, 20% yolk and 80% buffer solution gives best results. Penicillin and streptomycin should be added at the recommended rate. The yolk-citrate diluter has become the standard against which all new and modified diluters have been compared.

### 16-4.3
### Tris buffered-yolk

Tris buffered-yolk diluter is prepared by adding 20% fresh egg yolk to the tris buffer solution. Antibiotics are added at the recommended level. Six to 8% glycerol has been shown to be beneficial even when the semen is to be stored at 5°C. The tris buffer also offers an added advantage in that glycerol can be incorporated into the initial diluter, thus avoiding the time-consuming steps of adding glycerol at 5°C after the sperm have cooled. Most other diluters require this later procedure. Equilibration time does not seem to be as critical with tris-yolk diluter as with other diluters.

**16-4.4**
**Whole**
**homogenized**
**milk and skim**
**milk**

Whole homogenized milk and skim milk satisfy the requirements of a good semen diluter. Milk heated to normal pasteurization temperature contains a material, *lactenin,* which is spermicidal. Heating the milk to 90° to 95°C for 10 minutes inactivates lactenin. The only additions needed are the antibiotics to control microbial contaminants and glycerol if the semen is to be frozen. Whole milk has the disadvantage of poor sperm visibility under the microscope. This problem is apparently caused by light refraction by the fat globules contained in the whole homogenized milk.

**16-4.5**
**Other diluters**

The field of semen diluters has probably been more thoroughly researched since 1940 than any other research area of similar scope. The addition of many ingredients, such as simple sugars, amino acids, enzymes, and combinations of ingredients have been tried. Various concentrations of egg yolk and various mixtures of yolk-citrate and milk have been tried. Some of these combinations have been superior in some respects to the original yolk-citrate and most bull studs are using some version of the more complex semen diluter. Table 16-2 gives examples of the complex diluters compared to the yolk-citrate and milk diluters. The IVT and CUE diluters were developed for variable temperature use. Semen in IVT diluter was placed in glass ampules, flushed with $CO_2$, and sealed. Storage at ambient temperature showed some promise, but frozen semen replaced interest in variable temperature diluters.

TABLE 16-2. Simple and complex diluters satisfactory for use with bull semen

| Ingredients | Yolk-citrate | Homo-genized milk | Yolk-citrate-milk | I V T* | C U E# |
|---|---|---|---|---|---|
| Buffer (g/100 cc unless otherwise noted) | | | | | |
| Sodium bicarbonate | | | | .21 | .21 |
| Sodium citrate | 2.90 | | 2.90 | 2.00 | 1.45 |
| Potassium chloride | | | | .04 | .04 |
| Glucose | | | | .30 | .30 |
| Sulfanilamide | .30 | | .30 | .30 | .30 |
| Penicillin (IU/cc) | 1000 | 1000 | 1000 | 1000 | 1000 |
| Streptomycin ($\mu$/cc) | 1000 | 1000 | 1000 | 1000 | 1000 |
| Glycine | | | | | .937 |
| Egg yolk (% by volume) | 20 | | 15 | 10 | 20 |
| Milk | | 100 | 70 | | |
| Buffer (% by volume) | 80 | | 15 | 90 | 80 |

*Bartlett and VanDemark, J. Dairy Sci. 45:360, 1962.
#Foote *et al.*, J. Dairy Sci. 43:1330, 1960.

**16-5**
**Processing**
**bull semen**

The processing of semen involves dilution, cooling, packaging, and freezing. The average temperature of the semen as it reaches the processing room should be determined so that a water bath can be regulated to that temperature. The tube of freshly collected semen should be placed in the water bath while quality determinations are made or until processing can be initiated. Normally, processing is begun within 5 minutes. However, some research has indicated that holding undiluted semen of certain bulls at temperatures ranging from 26° to 32°C for intervals of 20 to 30 minutes improves fertility. These bulls are classified as problem bulls and the incubation temperature and time interval must be determined for each problem bull.

16-5.1
Semen dilution

The dilution of semen is carried out in two steps. The first step involves a predilution of the warm semen with three to four volumes of diluter for each volume of semen. The diluter used for predilution should be tempered in the same water bath used to maintain the temperature of the semen. The diluter used in this manner provides lecithin and lipoproteins to protect the sperm from cold shock during the cooling process. These materials apparently prevent changes in cell wall permeability during cooling. After the prediluted semen is cooled to 5°C it is diluted to final volume with diluter which has also been cooled to 5°C. The addition of glycerol will be discussed in a later paragraph.

16-5.2
Dilution rate

The main objective in deciding what the dilution rate should be is to provide the optimum number of motile sperm per breeding unit at the time of insemination. It is generally accepted that 10 million motile sperm at the time of insemination will provide optimum conception rate. When semen was used in the liquid form, determining the dilution rate was fairly simple. One needed to know the initial motility and sperm concentration of the ejaculate. The number of breeding units was determined by dividing the total number of motile sperm by 10 million. The procedure is a bit more complicated with frozen semen. Not only do we need to take initial motility and concentration into account, but we need to have some idea of the survival rate of the individual bull being processed. This can be determined only by freezing several ejaculates from a bull and making post-thaw evaluations on them. About 15 million motile sperm per breeding unit seems to be the minimum prefreeze concentration for ampules. Prefreeze concentrations of motile sperm as high as 30 million are not uncommon. Since the survival rate is greater for sperm frozen in straws than in ampules, fewer motile sperm are needed per straw unit. While 10 million motile sperm at the time of insemination is the accepted number, research indicates that this number can be reduced to 7 or 8 million for certain high fertility bulls without reducing conception rate. On the other hand conception rate may be enhanced for certain low fertility bulls by increasing the number of motile sperm to perhaps 15 million at the time of insemination. Accurate post-thaw evaluation is essential in determining the initial number of motile sperm required to obtain the desired number of motile or percent intact acrosomes at time of insemination.

**16-5.3 Cooling semen**

Cooling is accomplished by placing the prediluted warm semen container in a container of water at the same temperature (Figure 16-1). These are then placed in a refrigerator and cooled to 5°C. The combined volume of prediluted semen and surrounding water should vary according to the desired cooling rate. There is disagreement on the best rate of cooling for prediluted semen. Intervals ranging from 1 to 4 hours have been recommended. Most semen processors seem to be using cooling intervals ranging from 1¼ hours to 2 hours. Livability studies more consistently support the slower cooling rates, and several authors have shown that post-thaw motility favors slow cooling over a 2- to 4-hour period. It is probable that experiments which have shown no difference between fast and slow cooling rates on fertility have involved excessive numbers of motile sperm which mask the adverse effects of fast cooling.

Even though a cooling time of 2 to 4 hours appears most appropriate, cooling time may interact with initial holding time of semen at 30° to 35°C, equilibration time and with diluter composition. Each semen processor should know the rate of cooling which is optimum for their system. Continuous use of accurate thermometers and periodic use of thermocouples should be used to insure consistency.

FIGURE 16-1. Prediluted samples of semen are placed in water at the same temperature. The volume of water should be sufficient to allow cooling to 5°C in two to four hours.

**16-5.4 Glycerolation and equilibration**

Glycerol must be added to semen to protect it during freezing and thawing. Damage results from the selective freezing of free $H_2O$ both inside and outside the cells. This results in concentration of other cell constituents and solutes outside the cell. In addition to upsetting vital cell components, changes in permeability of cell membranes may occur. Preferential leakage of solutes and pH changes in nonelectrolytes have been mentioned as contributing to sperm damage.

Glycerol binds water and decreases the freezing point of solutions. Less ice is formed in its presence at any temperature. Concentration of solutes is correspondingly decreased. Glycerol partially dehydrates the sperm cells, further reducing the selective freezing of water. Other cryoprotective materials have been compared to glycerol. Dimethyl sulfoxide has a salt-buffering capacity and enters the cells rapidly. It is most effective with slow freezing. Glycols and a number of saccharides have shown some cryopreservative effects.

The level of glycerol varies somewhat with the diluter ingredients. Yolk-citrate and tris buffered-yolk diluters should contain 7% glycerol. Milk diluters containing 10 to 13% glycerol perform best. Glycerol must be added to semen diluted with yolk-citrate and milk diluters at 5°C. Sperm are damaged by glycerol in these diluters when added at 35°C.

*Equilibration time,* the time the sperm are allowed to remain in contact with the glycerol at 5°C, varies markedly between laboratories and with different diluters. From 4 to 18 hours have been recommended but most processors lean toward shorter periods. Some recent research indicates that the sperm reach equilibrium with the glycerol very rapidly (10 minutes or less) and that the benefit from several hours at 5°C results from further aging or maturing of the sperm prior to freezing. Additional research is needed for the final answer to this question.

The usual procedure for adding glycerol to semen involves dividing the diluter into two equal parts. Part 1 will contain no glycerol. To part 2, glycerol is added at twice the level desired in the final mixture. This would be 14% for yolk-citrate and tris-yolk diluters and 20 to 25% for milk diluters. Freshly collected semen is prediluted with part 1 diluter for cooling. After the semen is cooled, it is diluted to half the desired final volume with part 1 diluter. Part 2 diluter is added slowly over a period of at least one hour to bring to final desired volume. This glycerol portion can be dripped slowly into the part 1 portion or it may be divided into four equal parts and one part added to the part 1 portion at 15-minute intervals (Figure 16-2).

The glycerol can be added to the tris-yolk diluter at the time of preparation at the desired final level. The predilution and final dilution can be made with the tris diluter containing glycerol.

## 16-5.5
## Semen packaging: ampules and straws

Glass ampules with 0.5, 0.8, or 1.0 cc capacity were used almost exclusively for packaging from the onset of frozen semen until about 1970. The ampules were filled through a small opening and heat was used to melt the glass to form a seal (Figure 16-3). Six or eight ampules were clipped to an aluminum strip called a cane for storage in liquid nitrogen. The neck of the ampule was etched so that it could be broken to open the ampule.

FIGURE 16-2. One technique for dripping the diluter containing glycerol into the portion containing no glycerol. (Courtesy of Select Sires, Inc., Plain City, Ohio.)

FIGURE 16-3. Semen for freezing is packaged in either ampules (top) or plastic straws (bottom).

The 0.5 cc plastic straw has been the package of preference since about 1970. Plastic straws with 0.25 and 0.3 cc capacities are used in Europe but are not popular in the United States (Figure 16-2). One end of the straw contains a 3-part plug. A small amount of polyvinyl alcohol powder is placed between two small cotton plugs. The straws are filled by applying a vacuum to the end with the plug. The powder allows air to pass through as long as it remains dry, but when aqueous material (the semen) comes in contact with the powder it forms a seal which will not allow liquid or air to pass through. The opposite end of the straw is sealed either hydrostatically or ultrasonically after filling. The 0.5 cc straw is 113 mm long and 2.8 mm in diameter. One major advantage of the straw over the ampule is the conservation of storage space. Up to three times as many straws can be stored in a field or freezing unit as ampules. Most studies indicate that the straw offers the added advantage in increased sperm survival and a slight increase in conception rate over the ampule. Straws are placed in small plastic goblets which hold five or ten straws each. Two goblets are attached to an aluminum cane for storage in field units. The goblets, when filled with liquid nitrogen (LN), provide protection to the straws while being transferred from unit to unit.

Both the ampule and the straw must be labeled to identify the donor bull and the semen producing business. The full registration name and number of the bull plus his code number are printed on each unit. All semen producing businesses are assigned a code number by the U.S. Department of Agriculture and this number must also be printed on each unit.

Some recent research involving the freezing of semen in pellets offers some promise. Small depressions are made in the surface of a block of dry ice (Figure 16-7). Semen containing about ten times the usual number of motile sperm is placed in the depression to freeze. The usual volume is about 0.1 cc. After freezing the pellets are transferred directly to liquid nitrogen for storage. For insemination the pellets are thawed in enough warm diluter to provide adequate volume for insemination. Two major disadvantages of the pellet are identification of the individual pellet and microbial contamination incurred during handling. There is also some possibility of mixing sperm from different bulls from contaminated LN or forceps used to handle the pellets.

**16-5.6
Freezing**

The standard procedure for freezing is to place a single layer of straws on a tray. The tray is placed about 5.5 cm above the liquid nitrogen level of a large storage unit. Ampules are attached to canes and set vertically into baskets which are placed about 5.5 cm above the LN level. The cold nitrogen vapors in this area will freeze the semen at about the desired rate. Straws will reach the vapor temperature in about 2 minutes. The number of breeding units that can be frozen at one time and the number of batches that can be frozen during a working day depends a great deal on the size of the liquid nitrogen unit being used.

The reported research on rate of freezing indicates that there is a great deal of latitude. Table 16-3 summarizes some of the more recent studies. It appears that semen can be frozen both too fast and too slow. It would appear that rates between 126°C per minute down to 7°C per minute will give satisfactory results. It is logical to assume that the optimum rate of freezing might be influenced by several factors. Among these would be the type of package, glycerol level, recommended rate of thaw, and diluter composition.

TABLE 16-3.  Freeze rates of semen in straws* °C/minute

| Mortimer et al. 1976 from 5 to −130 | Robbins et al. 1976 from heat of fusion to −80 | Almquist and Wiggins 1973 | | Rodriguez et al. 1975 |
|---|---|---|---|---|
| | | 5 to −15 | −15 to −60 | |
| 135 | 126 | 82 | 43 | |
| 67 | 54 | 50 | 27 | |
| 38 | 30 | 43 | 25 | 38 |
| 19 | 15 | 32 | 23 | (5 to −130C) |
| 11 | 7.5 | | | 11.9 |
| 7 | | | | (10 to −130C) |
| | | | | 1.1 |
| | | | | (5 to −15C) |
| | | | | 3.9 |
| | | | | (−15 to −79C) |

*Dotted lines indicate freeze rates resulting in significantly depressed sperm viability post-thaw.
From Saacke, Proceedings 7th Tech. Conf. on A.I. and Reprod., 1978.

**16-6
Storage and
handling of
bull semen**

Semen that is going to be stored at above-freezing temperatures needs to be maintained at approximately 5°C. Frozen semen must be stored at a temperature below −75°C. The temperature of dry ice (−79°C) barely meets this requirement and one study showed a decline in NR rate of 13% between semen stored over 6 months compared to semen stored 1 to 2 months at −79°C. In addition to being difficult to handle and transport, dry ice must be replenished frequently. The lifespan of bull semen stored in liquid nitrogen has not been determined. Experiments using split ejaculates have shown no decrease in the NR rate for semen stored up to 2 years. One study reported services at 6-month intervals from 6 months to 5 years with no decrease in NR rates. Some calves have been born from semen stored up to 20 years, but the number of services are too few to draw conclusions.

Liquid nitrogen with a temperature of −196°C is the refrigerant of choice. Double wall stainless steel or aluminum containers with a vacuum between the walls make highly satisfactory storage units. Figure 16-4 shows a large storage unit and a small field unit. This very popular field unit has a nitrogen capacity of 20 liters and will hold for 90 days between charges. It will hold approximately 1200 straws of 0.5 cc capacity and approximately 600 ampules of 0.8 cc capacity. Most field units of this type are recharged at 60-day intervals to allow an adequate margin of safety.

Several precautions should be taken in handling semen. The packaged unit is very small, particularly the plastic straw, and when exposed to ambient temperatures of 25° to 30°C the temperature rises very rapidly. Repeated changes in temperature, particularly of temperatures above −79°C

FIGURE 16-4. Liquid nitrogen tanks used for frozen semen. A field unit is shown on the left and a larger storage unit on the right.

can be detrimental to the survival of the sperm. The top of the canister, shown in Figure 16-5, should remain 3 to 5 cm below the top of the nitrogen tank when looking for semen from a particular bull. The top of the cane should be labeled with the code number of the bull so that the individual straws or ampules will not have to be examined. Semen from more than one bull should never be placed on the same cane. Once the cane is identified, it should be withdrawn from the canister just far enough so that the straw or ampule can be removed without exposing the remainder of the units to ambient temperature. Specially designed forceps are available for lifting straws from the goblets. When semen is being transferred from one unit to another whole canes should be transferred insofar as possible so that the transfer can be made quickly.

It is a good idea to check the nitrogen level in field units about once a week. This can be done by inserting a ruler into the unit until it reaches bottom. Allow it to remain a few seconds before it is withdrawn and exposed to the atmosphere. Frost will form on the ruler showing the exact depth of nitrogen in the unit.

**16-7**
**Thawing**
**frozen bull**
**semen**

The optimum thawing temperature for frozen semen appears to be related to the rate of freezing. There probably are other interactions such as glycerol level, equilibration time and diluter composition. The effects of these interactions have not been worked out completely, but there is general agreement for thawing procedures.

16-7.1
Thawing
ampules

The geometric configuration of the ampule results in a freezing rate somewhat slower than that for the plastic straw. An ice water bath provides the proper rate of heat exchange to thaw the ampule for maximum survival rate (Figure 16-5). The ice water thaw is preferred to placing the ampule in the shirt pocket, allowing it to thaw at ambient temperature, or thawing in warm water.

Some precaution should be taken when thawing the ampule since occasionally an ampule will be improperly sealed. While immersed in liquid nitrogen over a long period of time, the nitrogen seeps into and fills the air space above the frozen semen. When the ampule and the liquid nitrogen are exposed to thawing temperatures the nitrogen expands faster than it can escape through the tiny opening and the ampule explodes. Therefore, the ice water used to thaw ampules of frozen semen should be in an unbreakable container such as a polystyrene thaw box or a soft plastic container.

**16-7.2**
**Thawing straws**

Straws should be thawed in a 32° to 35°C water bath (Figure 16-6). The actual thawing process will be completed in 12 to 15 seconds and the temperature of the semen will have reached approximately 5°C. Recent research shows increased conception rates when the semen remains in the thaw water for 30 seconds. Even when semen is used with ambient temperature below freezing, the 30-second thaw time gives better results. Some semen producing organizations recommend leaving the straw in the thaw bath for 45 to 60 seconds.

FIGURE 16-5. Ampules of frozen semen are thawed by transferring them to a container of ice water.

FIGURE 16-6. Straws of frozen semen are thawed by transferring them to a container of 32°C to 35°C water. Note thermometer in corner of thaw box.

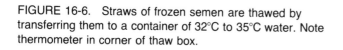

**16-8
Processing
boar, ram and
stallion semen**

The characteristics and properties of semen from these species require that different procedures be used in processing. AI for these species has been less successful, particularly with frozen semen.

**16-8.1
Processing boar
semen**

The semen from boars is low in concentration and high in volume. Dilution rates have necessarily been low, 1:1 to 1:5. Boar semen diluted 1:1 with a suitable diluter and cooled to 5°C in 2 hours will retain fertility for approximately 3 days. The dilute nature of boar semen has presented problems with freezing. Recent research at the Beltsville, Maryland Experiment Station seems to have at least partly overcome the problem. Aliquots of ejaculates containing $6 \times 10^9$ spermatozoa were centrifuged for 10 minutes at 300 g. The seminal plasma was poured off and the sperm were resuspended in 5 cc of Beltsville F5 (BF5) diluter (Table 16-4). The semen was cooled to 5°C over a 2-hour period. Five cc of BF5 diluter containing 2% glycerol was added and the semen was frozen immediately into pellets of 0.15 to 0.2 cc on dry ice (Figure 16-7). The pellets were transferred to liquid nitrogen for storage. Thawing was accomplished by removing 10 cc of pellets and holding them in a clean dry container for 3 minutes. The pellets were then placed in 25 cc of Beltsville thawing solution (BTS) (Table 16-4) at 50°C and swirled in the container until thawed. The thawed semen was poured

TABLE 16-4. Composition of Beltsville F5 diluter and thawing solution for boar semen

| Ingredient | Diluter Amount* | Thaw Solution Amount† |
|---|---|---|
| Tes-N-Tris (hydroxymethyl) methyl 2 aminoethane sulfonic acid | 1.2g | |
| Tris (hydroxymethyl) aminomethane | .2g | |
| Dextrose, anhydrous | 3.2g | 3.7g |
| Egg yolk | 20 cc | |
| Orvus ES Paste | .5cc | |
| Sodium citrate, dihydrate | | .6g |
| Sodium bicarbonate | | .125g |
| Sodium ethylenediamine tetraacetate | | .125g |
| Potassium chloride | | .075g |

*Brought to 100 cc with distilled water, centrifuged at 12,000g for 10 min. diluter decanted.
†Ingredients dissolved and brought to 100 cc with distilled water.
Purcel and Johnson, J. An. Sci. 40:99, 1975.

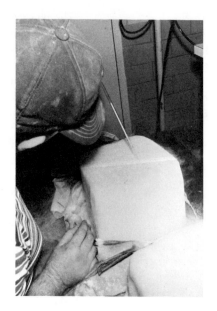

FIGURE 16-7. Boar semen is frozen in 0.1 to 0.15 cc pellets by pipetting the semen into small depressions in a block of dry ice. Pellets are placed in vials and stored in liquid nitrogen.

into an inseminating bottle. The thaw container was rinsed with 15 cc of BTS and the rinse solution was added to the inseminating bottle. This quantity represented a breeding unit. Eighty-five percent of 28 gilts produced fertilized ova to two inseminations and 87% of the ova produced were fertilized. The gilts were slaughtered 24 to 120 hours after the second insemination. The ova were recovered and examined for cleavage. Frozen boar semen is now available commercially.

**16-8.2
Processing ram
semen**

Contrary to what might logically be expected, ram semen does not react like bull semen. Ram semen may contain more than $3 \times 10^9$ sperm/cc, but is never diluted more than 1:10 and seldom more than 1:3. Sample ram semen diluters are shown in Table 16-5. The volume of an insemination unit ranges from 0.1 cc to 0.3 cc, but the number of motile sperm per unit has not been accurately defined. Total sperm numbers from $100 \times 10^6$ to $500 \times 10^6$ have been recommended. This large number, as compared to the number required for optimum conception in the cow, may be related to method of insemination. Insemination techniques are discussed in Chapter 17. In the ewe the semen is deposited in the posterior end of the cervix, but in the cow the semen is deposited in the anterior end of the cervix or in the body of the uterus. Conception is best with fresh semen (0 to 12 hours), but reasonable conception has been obtained with semen stored for 24 hours. Conception rates at 48 and 72 hours are unacceptable. Ram semen in buffered yolk diluter is stored at 5°C. There is evidence that 15°C is more favorable when skim milk diluter is used.

Ram semen has been frozen with moderate success, but the process is not considered practical at this time. The large number of sperm in a small volume is difficult to deal with when freezing semen. Several researchers have suggested that the time required for sperm to pass through the cervix may be a factor in the lowered conception for frozen semen which has somewhat lowered viability.

**16-8.3
Processing
stallion semen**

The recommended number of motile sperm per insemination of mares is $500 \times 10^6$. These can be suspended in 10 to 30 cc total volume with a satisfactory diluter. Table 16-6 shows some sample diluters. The cream-gel diluter gave significantly higher conception than the tris diluter. Conception with the cream-gel was comparable to that obtained with fresh undiluted semen. Yolk-citrate diluters have also been used.

Stallion semen can be cooled to 5° C and stored for 24 to 48 hours. Freezing has been accomplished with some success but not to the practical stage. More needs to be known about the physiology of stallion semen.

TABLE 16-5.   Sample diluters for ram semen

| Ingredients | Yolk-citrate | Skim milk |
|---|---|---|
| Sodium citrate dihydrate | 2g | |
| Glucose | 9.9g | |
| Egg yolk | 15cc | |
| Glycerol | 5cc | 5cc |
| Penicillin G (IU/cc) | 1000 | 1000 |
| Streptomycin ($\mu$/cc) | 1000 | 1000 |
| Heated skim milk | | 100cc |
| Distilled $H_2O$ to final vol. | 100cc | |

TABLE 16-6.   Sample stallion semen diluters

| Ingredients | Tris | Cream-Gelatin |
|---|---|---|
| | % | % |
| Tris (w/v) | 2.4 | |
| Glucose (anhydrous) (w/v) | .45 | |
| Citric acid (monohydrate) (w/v) | 1.255 | |
| Egg yolk (v/v) | 22.8 | |
| Glycerol (v/v) | 5.25 | |
| Half & half cream (v/v) | | 88.7 |
| Knox gelatin (w/v) | | 1.3 |
| Water (v/v) | 67.9 | 10 |
| Penicillin-Na-G (IU/cc) | 1000 | 1000 |
| Streptomycin-$SO_4$ ($\mu$/cc) | 1000 | 1000 |

From Pickett, et al., J. An. Sci. 40:1136, 1975.

**Suggested reading**

Bartlett, F. D. and N. L. VanDemark. 1962. Effect of diluent composition on survival and fertility of bovine spermatozoa stored in carbonated diluents. *J. Dairy Sci.*, 45:360.

Foote, R. H. and R. W. Bratton. 1950. The fertility of bovine semen in extenders containing sulfanilamide, penicillin, streptomycin and polymyxin. *J. Dairy Sci.*, 33:544.

Foote, R. H., L. C. Gray, D. C. Young and H. O. Dunn. 1960. Fertility of bull semen stored up to four days at 5°C in 20% egg yolk extenders. *J. Dairy Sci.*, 43:1330.

Pickett, B. W. and W. E. Berndtson. 1978. Principles and techniques of freezing spermatozoa. *Physiology of Reproduction and Artificial Insemination of Cattle.* (2nd ed.) ed. G. W. Salisbury, N. L. VanDemark and J. R. Lodge. W. H. Freeman and Co.

Pickett, B. W., L. D. Burwash, J. L. Voss and D. G. Black. 1975. Effect of seminal extenders on equine fertility. *J. Anim. Sci.*, 40:1136.

Polge, C. 1953. The storage of bull semen at low temperatures. *Vet. Rec.*, 65:557.

Purcel, V. G. 1979. Advances in preservation of swine spermatozoa. *Beltsville Symposia in Agricultural Research 3. Animal Reproduction.* Allanheld, Osmun and Co. Publisher; Allsted Press, a division of John Wiley & Sons.

Purcel, V. G. and L. A. Johnson. 1975. Freezing of boar spermatozoa: Freezing capacity with concentrated semen and a new thawing procedure. *J. Anim. Sci.*, 40:99.

Saake, R. G. 1978. Factors affecting spermatozoan viability from collection to use. *Proc. 7th Tech. Conf. Artif. Insem. Reprod.*, p. 3. NAAB, Columbia, Mo.

# Chapter

# 17 Insemination techniques

Successful AI culminates with the proper placement of high quality semen in the female reproductive tract. The object of insemination technique is to place the semen in the part of the reproductive tract that will give the best chances for conception. The insemination technique is different for each of the four farm species. This is due in part to size of the females, but also in part to the anatomy of the reproductive system.

**17-1**
**Insemination of the cow**

Three basically different methods of inseminating the cow have evolved since the beginning of AI.

17-1.1
Vaginal insemination

The earliest inseminations were accomplished by simply inserting a tube into the vagina and depositing semen at the mouth of the cervix. This procedure simulated a deposit of semen during natural mating and probably gave fair results when very large numbers of sperm were deposited. However, the environment of the vagina is not conducive to long life of sperm. Recent research with a limited number of services indicates that a modern breeding unit of semen containing approximately 10 million motile sperm will result in a very low conception rate when deposited in the vagina.

17-1.2
Cervical insemination

Cervical insemination is accomplished by inserting a sterile speculum (2 to 3 cm in diameter and 35 to 40 cm long) into the vagina. With the use of a light source (pen light or head lamp), an inseminating instrument can be inserted into the opening of the cervix (Figure 17-1). Normally the instrument can be inserted from 1 to 2 cm and the semen deposited at that point. This method is far superior to the vaginal method but usually gives 10 to 12 percentage units lower conception rate than the recto-vaginal method described below. Another disadvantage of this method is the amount of equipment that must be sterilized between inseminations.

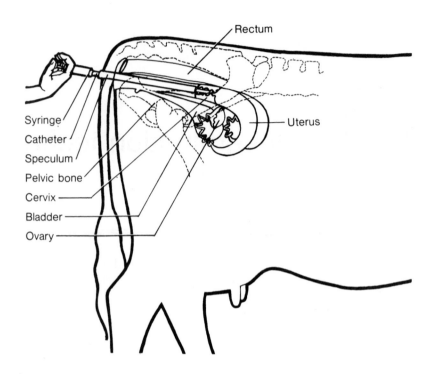

Rectum

Syringe
Catheter
Speculum
Pelvic bone
Cervix
Bladder
Ovary

Uterus

FIGURE 17-1. Speculum method for inseminating the cow. Similar method is used for the ewe.

**17-1.3
Recto-vaginal
insemination**

This method is also referred to as the *cervical fixation* method. It is accomplished by inserting a gloved left hand, lubricated with a small amount of surgical jelly, into the rectum of the cow. The hand locates and grasps the cervix (Figure 17-2). The cervix can be distinguished from the vagina and uterus by its firm, thick walls. The inseminating instrument is inserted through the vulva into the vagina until it contacts the cervix and the left hand. The lips of the vulva should be spread slightly when inserting the inseminating instrument to prevent contamination by the outer surfaces of the vulva. The cervix should be held by its posterior end with index and middle fingers and thumb, leaving the other two fingers free to help guide the inseminating instrument (Figure 17-3). The instrument is guided into the opening of the cervix and the left hand is used to thread the end of the instrument through the irregular cervical channel. The cervical folds make it necessary to manipulate the cervix in all directions in order to pass the instrument through the cervix. As the instrument progresses through the cervix, the fingers and thumb are moved forward so that the manipulation is taking place just forward to the end of the inseminating instrument. Progress of the instrument can be determined by the rigidity it gives to the cervix.

The student is referred to Chapter 2, Section 2-4 for the anatomy of the cervix. The inseminating instrument should be stopped as soon as it reaches the end of the cervix.

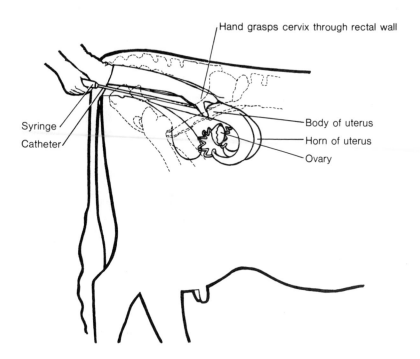

FIGURE 17-2.   Recto-vaginal method for inseminating the cow.

The inseminating instrument should not be removed from the vagina after insertion until insemination is completed. Extra passages through the lips of the vulva result in greater contamination. It is particularly important not to withdraw the instrument when the cow urinates. The opening would be pointed directly into the oncoming stream of urine, which is extremely detrimental to sperm.

The beginner should be aware of some problem situations.

1.   The insemination instrument should be inserted into the vagina with the forward tip held higher than the other end. This will help prevent the insemination instrument from entering the suburethral diverticulum or the external urethral orifice.

2.   Occasionally, muscular contractions will force the reproductive tract toward the anus causing the vagina to become folded. This may make it

FIGURE 17-3. Right way (top) and wrong way (bottom) of holding the cervix for the recto-vaginal method of insemination.

seemingly impossible to bring the inseminating instrument into contact with the cervix. The cervix can be grasped with the left hand and pushed forward to straighten the vagina.

3. The cow will attempt to expel the left hand from the rectum with *peristaltic* muscular contractions. The contractions begin at the junction of the large intestine and the rectum and proceed toward the anus. When the contraction reaches the hand, its progress is stopped but the muscles continue to squeeze the hand. The hand will tire quickly if the contraction is fought. The cervix must be released and the hand pushed through the contraction. The rectum will be relaxed so the cervix can be manipulated again.

4. The rectal muscles may contract, forming a large, hard walled, tubular structure. The cervix cannot be felt or manipulated through this condition. This contraction can be overcome by reaching forward to the junction of the rectum and large intestine. The fingers can be cupped over the hardened rectal wall and the hand pulled toward the anus. This procedure usually causes the contracted rectal muscle to relax and soften so the cervix can be manipulated.

5. In practice situations the vagina occasionally fills with air making it difficult to grasp and manipulate the cervix. Firm pressure with the hand toward the vulva will dispel the air and restore favorable working conditions.

6. An extremely full bladder can make working with the cervix difficult. Manipulating the clitoris may cause urination.

The recto-vaginal method is more difficult to learn. However, its superior conception rate makes it the method of choice. Conception rates are lower for the beginner, but as the technique is mastered conception improves.

## 17-1.4 Inseminating equipment

A plastic inseminating catheter fitted with a polypropylene bulb (poly bulb) or a 2-cc syringe attached to the catheter with a short rubber connecting tube must be used when liquid semen or semen frozen in ampules is being used (Figure 17-4). Care should be taken to draw all of the semen from the ampule and this will require some practice when the poly bulb is used. Care must also be taken not to draw the semen into the syringe when one is used. The semen should be discharged from the inseminating catheter slowly. If an attempt is made to expel the semen quickly the surface tension between the semen and the inseminating catheter is such that the air will be forced through the semen column leaving 25 to 50% of the semen in the catheter.

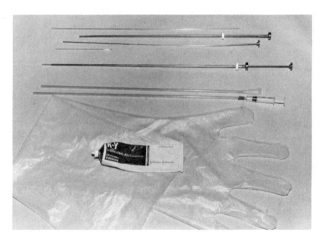

FIGURE 17-4. Equipment used to inseminate the cow. *Bottom:* Plastic glove and KY lubricating jelly. *Next:* Plastic catheters with 2 cc syringe and poly bulb. *Next:* Assembled half cc straw gun. *Top:* Unassembled straw gun; straw, plunger, barrel, and outer plastic sheath.

Insemination using semen packaged in the plastic straw requires specially designed equipment. The "straw gun" is a stainless steel tube with a small stainless steel rod to serve as a plunger (Figure 17-4). The lumen of the tube is the same diameter as the lumen of the straw except for a chamber in the anterior end, which is large enough to accommodate the exterior diameter of the straw. This chamber is one cm shorter than the straw so that the straw protrudes beyond the end of the gun. The straw is inserted with the plug seal going in first. The electrostatically sealed end is cut perpendicular to the straw with surgical scissors. A plastic sheath is slipped over the end of the straw and the gun. The posterior end of the gun tapers so that a plastic O-ring can be used to hold the sheath in place. The sheath is designed to fit tightly against the end of the straw forming a seal to prevent the semen from being trapped between the straw and the sheath. The semen is expelled by pushing the stainless steel rod against the straw plug, which then acts as a plunger to deliver the semen to the site of insemination. This positive action insures maximum delivery of sperm.

**17-1.5
Site of semen
deposition**

The final chapter on the ideal site for depositing semen in the cow possibly has not been written. Many experiments have compared conception rates for different sites of deposition. Vaginal and posterior cervical deposition are definitely inferior to midcervical, body of uterus or deep uterine depositions. Most research shows little difference between body of the uterus and midcervical; however, no experiments have favored midcervical. Deep uterine horn deposition is not recommended because of possible trauma and/or infection.

The cervix has been shown to provide a more favorable environment for sperm. There is some evidence to support the thesis that semen deposited in the cervix will form a pool. Sperm from this pool move out gradually over a period of time, possibly providing viable sperm at the site of fertilization for a longer period than when semen is placed in the uterus. There is little difference in the transport time to the site of fertilization when semen is deposited in midcervix or body of the uterus. It has been suggested that the same situation may exist when semen is deposited at the cervical-uterine junction. A small amount of the semen may be drawn back into the cervix and discharged as described above. More research is needed in this area.

Assuming no differrence in conception rate, midcervical deposition has the following advantages: (1) the reduced danger of damaging the uterine wall; (2) the utilization of the antibacterial action of cervical secretions; and (3) the decreased danger of interrupting pregnancy with repeat breedings.

Other research has shown that the conception rate of cows in which the semen was deposited in midcervix because the inseminator was unable to pass the insemination instrument beyond that point, was about 15 percentage units lower than for cows in which the semen was deposited in the body of the uterus. This reduced fertility may have been due to irritation and swelling of the cervical mucosa or other factors which prevented the semen from traversing the cervix.

**17-2
Insemination
of the ewe**

Because of the size of the ewe, it is necessary to use either the vaginal or cervical method described for the cow. A small stainless steel spreading speculum is usually used on the ewe. With the aid of a head light, the inseminating instrument is inserted into the mouth of the cervix where the semen is deposited (Figure 17-5). The insemination unit contains $500 \times 10^6$ motile sperm in a small volume such as the 0.25 cc straw.

FIGURE 17-5. Inseminating the ewe. A stainless steel spreading speculum is used. (Courtesy of E. K. Inskeep. Division of Animal and Veterinary Sciences, West Virginia University.)

**17-3
Insemination
of the sow**

The anatomy of the cervix in the sow makes insemination easy. The inseminating tube can be inserted into the cervix without the aid of sight or cervical fixation. A special inseminating tube with a design similar to the boar's penis has been developed for inseminating the sow (Figure 17-6). The instrument is available commercially. The anterior end is designed to fit into the cervix so that it simulates the locking of the boar's penis into the cervix during natural mating.

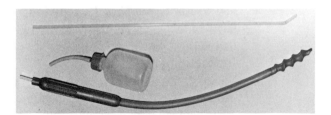

FIGURE 17-6. Equipment used to inseminate the sow. *Top:* Plastic insemination catheter bent at a 30° angle two cm from the end. *Center:* Plastic squeeze bottle used to force semen through either catheter. *Bottom:* Nasco's plastic spirette, an imitation of the boar's penis.

Another technique involves heating a plastic inseminating tube to form a 30-degree bend about 2 cm from the end of the tube (Figure 17-6). This bent tube is inserted into the opening of the cervix and turned so that it threads its way into the corkscrew channel of the cervix. It also allows the end of the catheter to bypass the external urethral orifice without danger of entering the bladder instead of the cervix. A plastic squeeze bottle or a large syringe is used to express about 50 cc of semen into the cervix.

**17-4**
**Insemination**
**of the mare**

The method used for inseminating mares requires special emphasis on cleanliness, because the hand is placed in the vagina with a finger through the cervix. After confining the mare in a chute suitable to protect the technician, her tail is wrapped with cheesecloth and tied to one side. The vulva, anus, and surrounding area are then scrubbed with water and mild soap, with special emphasis on cleaning the creased areas on either side of the vulva. After rinsing with water, the area is dried with sterile gauze. A shoulder length plastic glove is used by the technician with a sterile surgeons' glove worn over the plastic glove. A 50 cc syringe with a volume of semen containing $500 \times 10^6$ motile sperm is connected to a plastic inseminating catheter. The gloved hand is placed in the vagina with a finger through the cervix, the catheter is passed beside the hand and through the cervix and the semen is deposited in the body of the uterus. It is easy to pass the catheter through the mare's cervix because it is dilated and softened during estrus. (See Figure 17-7.)

**Suggested**
**reading**

Foote, R. H. 1974. Artificial insemination. *Reproduction in Farm Animals.* (3rd ed.) ed. E. S. E. Hafez. Lea and Febiger.

Holt, A. F. 1946. Comparison between intracervical and intrauterine methods of artificial insemination. *Vet. Rec.*, 58:309.

Salamon, S. and R. J. Lightfoot. 1970. Fertility of ram spermatozoa frozen by the pellet method. III. The effects of insemination technique, oxytocin and relaxin on lambing. *J. Reprod. Fert.*, 22:409.

Salisbury, G. W. and N. L. VanDemark. 1951. The effect of cervical, uterine and cornual insemination on fertility of the dairy cow. *J. Dairy Sci.*, 34:68.

Sullivan, J. J., D. E. Bartlett, F. I. Elliott, J. R. Brovwer and F. B. Kloch. 1972. A comparison of recto-vaginal, vaginal, and speculum approaches for insemination of cows and heifers. *A.I. Digest*, 20:6.

Wilcox, C. J. and K. D. Pfav. 1958. Effect of two services during estrus on the conception rate of dairy cows. *J. Dairy Sci.*, 41:997.

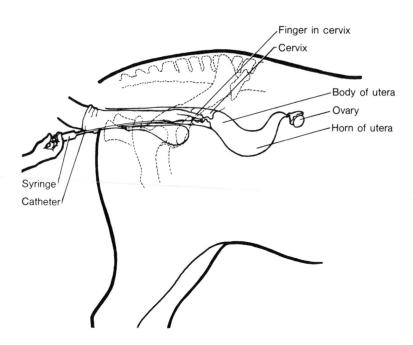

FIGURE 17-7.   Method used to inseminate the mare.

# Management for improved reproduction

# Chapter

# 18

# Altering reproductive processes

A great deal of progress has been made during the past 30 years through better management of the natural reproductive processes. However, livestock managers and researchers have recognized that certain natural reproductive processes can be altered to the advantage of management. Artificial insemination is an example of how tremendous improvement can be made in both genetic and reproductive management of livestock. AI has been adequately discussed in previous chapters so this chapter will be devoted to other means of altering reproductive processes to enhance livestock management.

## 18-1 Estrus synchronization

Several advantages have been recognized in having a number of females in estrus during a very short period of time. It will permit the manager to schedule livestock handling and breeding to fit into a work schedule with other required activities. The time-consuming job of estrus detection might be eliminated. The breeding season can be shortened and more females become pregnant during the first week of breeding. Animals can be grouped into desired parturition patterns so that intensive care may be provided for limited periods. Likewise, parturition season can be shifted to better coincide with the most favorable marketing patterns.

Estrus synchronization would make AI much more attractive to the commerical beef cattle manager by making it possible for him to get all of his cattle inseminated within one week. This is possible by handling the cows only twice, once for treatment and once for insemination. Having a high concentration of dairy heifers entering the milking string at one time has both advantages and disadvantages. Since heifers tend to have more difficulty calving than cows, having several calve during a short period would require less observation time. On the other hand, heifers require extra time and effort in the milking routine. Many dairy herd managers do not want to have a large number of heifers to "break in" at the same time.

**18-1.1
Progestin
method**

Early work demonstrated that injections of progesterone inhibited estrus and ovulation in cattle and sheep. The period of administration must be sufficiently long to allow the corpora lutea to regress in order to obtain synchronization. The treatment period was usually 16 days for cows. The exogenous progestin prevents the release of FSH so that estrus and ovulation are prevented until progestin is withdrawn. Upon withdrawal of progestin the decreasing blood level of progestin results in FSH release and estrus occurs 2 to 6 days later.

The next step in estrus synchronization was the testing of synthetic progestins on cows, ewes, and sows. The progestins were administered in feed, in drinking water, as subcutaneous implants, topical application, and vaginal pessaries. Estrus occurred over a 4-day period beginning 2 days after progestin withdrawal. About 80 to 90% of the females were synchronized. Some showed conception in synchronized females to be equal to controls, but overall the conception rates are at least 15% lower following progestin treatment. Figure 18-1 shows a subcutaneous implant of progestin being made.

The need for more acceptable conception rates and better synchronization prompted progestin treatment to be combined with either estrogen or gonadotrophin injections. Conception rates of 69% have been reported for cycling ewes treated for 14 days with vaginal pessaries containing fluorogesterone acetate and a single injection of 400 IU of pregnant mare's serum gonadotrophin at a time of pessary removal. The use of estradiol valerate

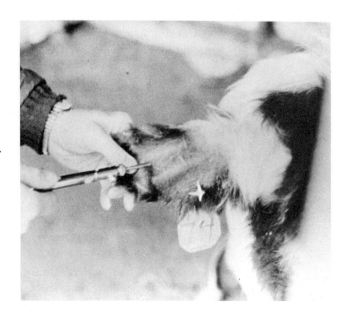

FIGURE 18-1. A progestin implant being inserted into a cow's ear. (Courtesy of W. Humphrey, Anim. Sci. Dept., Miss. State Univ.)

injections at the time of placing progestin implants in beef cattle resulted in conception rates equal to or greater than controls and improved synchronization. Estradiol speeds luteolysis and shortens the progestin treatment period.

Synchronized females may be inseminated at observed estrus or may be time-inseminiated. Some researchers have reported that two properly timed inseminations based on progestin withdrawal results in more pregnant cows than breeding at detected estrus. Some treated cows ovulate without showing overt signs of estrus and conceive to the timed-inseminations. However, the conception rate is higher for the cows bred during detected estrus. The data in Table 18-1 show the conception of heifers time-inseminated following progestin treatment. One timed-insemination 54 hours after implant removal gave results comparable to two timed-inseminations at 48 and 60 hours after implant removal. Seventy-two hours after implant removal is too late to inseminate and 48 hours may be too early.

A more practical approach may be to combine estrus detection with timed-insemination. All females not detected in estrus could be inseminated after a predetermined time. When expensive semen is being used, fewer units would be required, but the silent ovulators would still have a chance to conceive. Good managers may want to go a step further and palpate those cows not showing signs of estrus for the presence of a follicle before breeding. Some horse breeders are following this practice for timing inseminations.

TABLE 18-1.   Effect of timing of insemination on the pregnancy rate of heifers treated with Norgestomet ear implants for 9 days*

| AI after implant removal | No. of heifers | % pregnant |
|---|---|---|
| 48 & 60 hrs | 148 | 66.2 |
| 48 & 72 hrs | 145 | 62.1 |
| 54 hrs | 137 | 65.7 |
| Total | 430 | 64.7 |

*When the 6 mg implant was inserted 5 mg of estradiol and 3 mg of Norgestomet were injected IM.
From Wishart and Drew, *Vet. Record* 101:230, 1977.

**18-1.2
Prostaglandin
method**

Prostaglandin F$_2\alpha$ administered on day 5 through day 17 will cause a regression of the corpus luteum and a subsequent return to estrus in 36 to 72 hours. The day 1 through day 4 corpus luteum apparently does not have sufficient receptor sites to respond to normal levels of PGF$_2\alpha$. Some response to very large doses has been observed. Metabolic pathways may not have developed sufficiently at this point to allow complete response. After day 17 the corpus luteum is normally regressing.

When protaglandin is injected intramuscularly (IM), 25 to 40 mg are required to regress the CL. Five mg placed in the uterus will achieve the same results. When administered IM, the prostaglandin passes through the lungs where a high percentage is deactivated prior to reaching the ovary. Whereas that placed in the uterus travels directly to the ovary by way of counter current circulation (Chapter 2), thus requiring a much smaller dosage. Only 0.5 mg of the prostaglandin analog injected intramuscularly is required to regress the CL. This material is not deactivated by the lungs.

$PGF_{2\alpha}$ or the prostaglandin analog administered as described in the preceding paragraph between day 5 and day 17 of the estrous cycle has caused blood plasma progesterone levels to fall rapidly within 24 hours, followed by a rise in estrogen within 24 hours. An ovulatory peak of LH occurred on the average within 3 days with estrus occurring at about the time of the LH peak. Ovulation occurred about 24 hours after the onset of estrus.

Since recycling of cows with prostaglandin is only effective when administered to cycling cows between day 5 and day 17 of the cycle, one can only expect about 60 to 65% of the cows to respond at any given time. The herd manager then is faced with four alternatives:

1. Rectally palpate all cows and inject only those that are judged to have CL's that are 5 days or more in age. Treated cows may be time-inseminated 80 hours following treatment or inseminated during detected estrus. Again timed-insemination may result in lower conception rate based on total cows bred. The remaining cows could be treated 12 days later and inseminated as the first group.

2. Treat all cows, inseminate those that come into estrus within 5 days, and then retreat all remaining cows 12 days following the first treatment. Cows treated the second time may be time-inseminated at 80 hours post-treatment or inseminated during detected estrus.

3. Check estrus for 5 days and inseminate those observed in estrus; treat remaining animals with $PGF_{2\alpha}$ on day 5 and inseminate at the ensuing estrus. This method has resulted in 65 to 70% conception rate.

4. Treat all cows, wait 12 days, then retreat all cows and either time-inseminate at 80 hours or inseminate during detected estrus. It must be remembered that only cycling cows will respond to prostaglandin treatment.

Conception rates following prostaglandin treatments have generally been comparable to those obtained at naturally occurring estrus. A much higher percentage of cows become pregnant during the first 5 days of breeding whether synchronized with progestins or with prostaglandins than is possible with heat detection of unsynchronized cows. The unsynchronized cows had estrus distributed over a 21-day period. This advantage is still significant at 25 days of breeding, but disappears by day 45. Thus, synchronization reduces time and effort of heat detection and artificially inseminating cows and results in a higher percentage of the herd becoming pregnant early in the breeding season.

## 18-1.3 Combination of progestins and prostaglandins

A combination of the two methods has shown a great deal of promise. Progestin administered as a vaginal pessary for 7 days combined with $PGF_2\alpha$ administered on day 6 resulted in conception equal to or better than controls with Holstein heifers. Synchronization was also improved by the technique; 82% of treated animals exhibited estrus within a 17 hour period and 100% within a 32 hour period. The combination treatment has some advantages: (1) shortens the period of progestin treatment and thus possibly enhancing the chances of conception; (2) requires only one prostaglandin treatment; (3) shortens the overall synchronization; and (4) gives better synchronization.

## 18-1.4 Estrus synchronization in mares

$PGF_2\alpha$ has been effective in synchronizing estrus in mares starting on day 6 post-ovulation through day 18 of the cycle. After treatment they return to estrus in 4 to 5 days and ovulate 10 to 12 days post-injection. When injected on day 6, 80 to 90% of the mares respond. It is necessary to know the status of the ovary with respect to an active corpus luteum or a follicle before treatment.

The livestock manager should be fully aware that the synchronization of estrus is not a substitute for a good herd management. Research conducted in privately owned herds has shown that synchronization can be highly successful in well managed herds and a dismal failure in herds with only marginal management. Use only those drugs that have cleared FDA for specific classes and ages of livestock.

## 18-2 Treatment of anestrus

Anestrus is a condition in which the ovaries are inactive. It is a normal phenomenon during the sexually inactive period of seasonal breeding species. It is also a normal condition during the postpartum interval of cattle. In well-managed dairy herds, cows experience anestrus for 30 to 40 days postpartum. However, even in well-managed beef herds anestrus may persist for a considerably longer period of time. It is not uncommon for the average first estrous period for beef herds to occur at 120 days postpartum. The major difference between dairy and beef cattle would appear to be the calf suckling the beef cow. Marginal management which contributes to low energy intake obviously aggravates the situation.

## 18-2.1 Shang treatment

This treatment involves an ear implant of progestin plus an intramuscular injection of progestin and estradiol valerate at the time of implant. After 9 days the implant is removed and the calves are separated from their dams for 48 hours. The cows may be time-inseminated at 48 to 54 hours after removal of the implants or they may be inseminated during observed estrus. This system offers all of the advantages of estrus synchronization described in the preceding section plus the separation of the cow and calf terminates the anestrus condition in many cows. Again, a large field trial has shown the treatment to be about twice as effective in well-managed herds as it is in marginally-managed herds. The treatment should not be administered until the cows are approximately 60 days postpartum.

**18-2.2**
**Creep feeding**
**of calves**

Studies with both progestins and prostaglandins have shown better responses when the calves of the cows to be treated had access to creep feed for several weeks prior to treatment (Figure 18-2). Having access to the feed apparently reduces the frequency of suckling, thus, making it easier to break the period of anestrus. A combination of creep feeding followed by the shang treatment might give better results than either used alone.

FIGURE 18-2. Calves allowed access to a creep feeder tend to nurse less often.

**18-3**
**Super-**
**ovulation and**
**ova transfer**

Artificial insemination makes it possible to utilize the sperm produced by the male reasonably efficiently. The potential ova contained in the ovary of the female have not been utilized effectively. Superovulation is a procedure in which the female is treated with hormones to cause her to produce several ova instead of the one that she normally produces at each estrus. In the ewe or cow, an average of 12 ovulations can be expected. In one study with 40 donor cows an average of 7.9 ova were collected nonsurgically. From 60 to 70% of superovulated ova are normal embryos. Reliable superovulation has not yet been achieved in the mare, but can be accomplished in sows.

Superovulation in calves has been accomplished and when made practical will shorten the generation interval by a year or more. Superior germ plasm could be identified earlier and utilized more. To date the response has been variable and fertilization rates have been very low.

18-3.1
Techniques of
superovulation

The cow and the ewe normally produce only one ovum per estrous period, while the sow produces 12 to 20 ova. The object of superovulation is to cause the female to produce a large number of ova that can be transferred to other females.

**Ewes**  Six hundred to 1000 IU of pregnant mare's serum gondotropin (PMSG) injected subcutaneously on day 12 or 13 of the estrous cycle has been effective with ewes. The level of LH released from the ewe's pituitary gland apparently is adequate to cause ovulation of the induced follicles without the injection of additional LH.

**Cows**  Two thousand to 2500 IU of PMSG injected during mid-luteal phase of the cycle followed in 2 or 3 days with an injection of $PGF_{2\alpha}$ is the most common procedure for superovulating cattle. Mature cows do not require exogenous LH. Mid-cycle luteolysis yields more embryos than initiating treatment on day 15 to 18 of the estrous cycle. This treatment usually causes twelve or more ovulations (Figure 18-3) but may only yield four to six fertilized ova when the treated cow is inseminated with multiple doses of semen. Perpuberal heifers show even greater response to superovulation treatment than cows. However, the fertilization rate of the ova produced has been extremely low. Exogenous LH or HCG is required for ovulation.

**Sows**  In sows 750 to 1500 IU of PMSG (dose varies with size) should be injected subcutaneously on day 15 of the cycle. At the onset of estrus 500 IU of HCG is injected intravenously.

FIGURE 18-3.  Multiple ovulation sites can be seen on this superovulated ovary which has been exposed by surgery. (Courtesy of Codding Embryological Services, Foraker, Okla.)

Theoretically superovulation can be accomplished as often as the length of the estrous cycle. With mid-luteal initiation of treatment the interval may be shortened. There are still some problems which must be solved related to repeated superovulation treatments. Extreme variation between animals and diminished response with repeated treatments are problems. Assuming that these problems can be overcome, up to 5 fertilized ova from a cow might be obtained every 15 days. This would mean 120 or more fertilized ova from one cow in a 12-month period. The reduced response may be due to an immune reaction in which antibodies are produced in response to the injections of a foreign protein. It may not be solved easily.

**18-3.2**
**Ova collection**

Early collection techniques involved either slaughtering the females and excising the oviducts or surgically removing the oviducts from the live females at 72 hours post-ovulation so that the ova could be recovered by flushing. This defeated the primary purpose of superovulation so other methods were developed. A surgical method was developed first and is accomplished by performing a *laparotomy* (flank or midline abdominal incision) to expose the reproductive tract (Figure 18-4). A clamp is placed near the tip of the uterine horn so that fluid can be forced from the tip of the uterine horn through the oviduct toward the ovary. This fluid carries the ova with it and is collected at the infundibulum. The procedure allows for the recovery of a high percentage of ova, but because of the surgical trauma and resulting adhesions it can be repeated only a few times. The adhesions make it difficult, if not impossible, to expose the reproductive tract repeatedly.

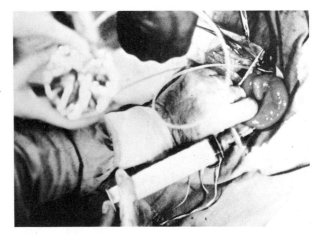

FIGURE 18-4. A surgical procedure is being used to flush ova from the oviduct. (Courtesy of Codding Embryological Services, Foraker, Okla.)

Nonsurgical techniques of recovery have been developed for the cow and mare that give results essentially equal to surgical methods. They involve the use of a Foley catheter (two-way flow catheter) which allows flushing fluids to pass into the uterus and at the same time allows fluids to be returned from the uterus to a collecting receptacle (Figure 18-5). A small balloon near the end of the catheter, which can be inflated just inside the cervix to prevent the flushing fluid from escaping through the cervix, is also a feature. With this method it is essentially impossible to determine how many ovulation sites are present on the ovary, so it is not possible to determine when all of the ova have been collected. In controlled experiments about 50% of superovulated ova are recovered by either surgical or nonsurgical procedures.

The fluids used for flushing the ova must be compatible with the particular species involved. For the ewe, homologous blood serum alone or with an equal volume of sterile 0.9% saline should be used. The serum must be heated to 55°C for 30 minutes to deactivate any ovicidal factor. For cows and sows, a commercial tissue culture medium can be used. The addition of 1000 units of penicillin and 500 to 1000 $\mu$g of streptomycin per cc of medium is necessary to reduce the chances of infection in the donor cow as well as the transmission of infection to the recipient.

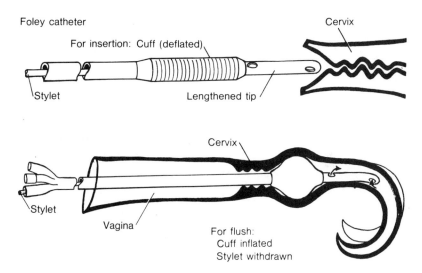

FIGURE 18-5.  Foley catheter used to flush ova from the uterus by nonsurgical means.

**18-3.3
Ova storage
and transfer**

Ova should be maintained at near body temperature in the media used for flushing during the period between recovery and transfer to the recipients. Ova may be held in this manner up to 10 hours without a reduction in survival. Lower temperature must be used if ova are to be retained for longer periods of time. Sheep ova held at 10°C have retained viability up to 72 hours (Figure 18-6). Embryos from cows and ewes may be frozen and stored in liquid nitrogen but a 50% loss of viability should be expected.

Fertilized ova must be transferred to recipients that have had their estrous cycles synchronized with the donor. The synchronization techniques have been discussed in earlier sections of this chapter. Both surgical and nonsurgical techniques have been used for transferring ova. The surgical procedure involves a laparotomy to expose the reproductive tract. The normal fertilized ovum or ova, depending on species, is placed either in the oviduct or the uterus of the recipient. A small syringe fitted with a 21-gauge needle is used to make the transfer. When the ovum is placed in the uterus, the needle is carefully inserted through the wall of the uterine horn adjacent to the ovary which contains the corpus luteum. When the ovum is placed in the oviduct the needle is carefully inserted through the infundibulum into the ampulla where the ovum is deposited.

Two nonsurgical procedures for transferring ova to recipients have been described and used in the cow and mare. In one procedure a long hypodermic needle is inserted into the vagina and used to bypass the cervix and puncture the uterine horn so that the egg can be deposited into the lumen of the uterus. For the second procedure, the ovum is placed in a plastic straw which has been described for packaging semen. This straw is then placed in the stainless steel inseminating gun which is passed through the cervix for the deposit of the ovum in the uterus. Thus far, the nonsurgical technique has not given as consistent results as surgical methods but offers a great deal of promise.

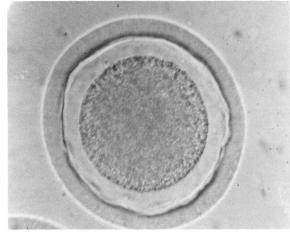

FIGURE 18-6. An unfertilized sheep ovum.

**18-3.4
Promoting
twinning**

Modifications of the superovulation treatment discussed in Section 18-3.1 have been employed to induce twinning in cattle and in sheep. While the number of offspring per pregnancy can be increased it is not possible to limit the number of ovulations to only two. When the treatment produces several ova, space limitations in the uterus usually result in embryonic mortality to reduce the number of feti to one, two, or three. Unfortunately, it does not always occur this way, but instead too many implantations frequently occur and later result in abortions.

While twinning is desirable in sheep, the disadvantages outweigh the advantages for twinning in cattle. Some of these are: (1) short gestations accompanied by small, weak calves; (2) high incidence of retained placentae;

(3) more stress on the cows—dairy cows produce 10% less milk in the lactation following the delivery of twins; (4) increased reproductive problems—one study reported that 50% of the cows producing twins failed to calve the next year; (5) increased incidence of cystic ovaries—anytime exogenous FSH, particularly in the form of PMSG, is used one must expect an increase in the incidence of cystic ovaries; and (6) most beef cows do not produce enough milk for two calves.

## 18-4
## Induced
## parturition
### 18-4.1
### Advantages and disadvantages

Livestock managers have indicated a need for a safe and effective means of inducing parturition. Some possible advantages are: (1) to group parturitions that would normally be spread over a 1 or 2 week period into 24 hours to facilitate closer observation; (2) to have more offspring of a uniform age and size at time of marketing; and (3) to shorten gestation and generation interval. The benefits to be derived from (2) and (3) are very small, so realistically there is only one major advantage for induced parturition.

At the present time, the disadvantages far outweigh the advantages. In sheep the treatments for inducing parturition are only effective when administered very near to term. This requires that the exact breeding dates be known and even then only very small groupings can be accomplished. In cattle the incidence of retained placentae is very high following induced parturition. In some studies more than 50% of treated cows retained their placentae. One study utilizing a small number of cows indicated that an injection of 6 mg of estradiol benzoate IM at the time of induction treatment reduced the incidence of retained placentae. Another more complete study showed that estradiol-17$\beta$, estradiol-17$\alpha$ and estrone were not effective in reducing the incidence of retained placentae.

### 18-4.2
### Techniques

It has been established that the fetus initiates parturition in sheep and that an intact hypothalamic-pituitary-adrenal axis in the fetus is required. This and other data have led scientists to accept fetal cortisol as the compound which triggers parturition. The student is referred to Section 9-3 for further information on parturition.

Induction of parturition has been accomplished in both cattle and sheep by IM injections of dexamethasone and other corticoids into the dam. PGF$_{2\alpha}$ and its more potent analog will also induce parturition in these species. A thorough understanding of the endocrinology of parturition will help one visualize means by which parturition may be induced.

## 18-5
## Control of sex

Researchers have tried for more than 50 years to separate the X and Y chromosome bearing sperm with little success. Centrifugation, sedimentation, electrophoresis, selective killing, pressure changes, pH changes, and other techniques have been used. Some changes in sex ratios of laboratory animals have been accomplished but essentially no progress has been made with farm animals.

Recent results with human semen offers considerable hope. A staining technique using fluorochrome quinacrine will selectively stain the Y chromosome. This may make it possible to determine the effectiveness of separation research without having to use the semen to produce offspring for sex ratio determination. So far the technique does not work on the sperm of other species.

Other work with human semen utilizing varying densities of bovine serum albumin (BSA) has demonstrated considerable progress in separating the X and Y bearing sperm (Table 18-2). These researchers suggested that the X bearing sperm is slightly larger and heavier than the Y bearing sperm, and the Y sperm has the ability to swim faster and more readily through viscous media. These two factors may have contributed to the separation success.

TABLE 18-2.   Y sperm yield with a three step isolation process

| Fractions | % Y Sperm | % Sperm motility | % Sperm recovery |
|---|---|---|---|
| Isolation fraction #1 (6% BSA) | 64 | 85 | 26 |
| Isolation fraction #2 (10% BSA) | 68 | 95 | 52 |
| Isolation fraction #3 (20% BSA) | 85 | 98 | 44 |

Adapted from Ericsson, et al. Nature, Vol. 246: 421, 1973.

**18-6**
**Altering male**
**reproduction**

There are two practical reasons for altering male reproduction. The most common, castration, is performed for the purpose of producing a higher quality carcass in meat animals. Other alterations have been performed to produce males that can be used to assist in estrus detection.

**18-6.1**
**Castration**

Castration involves surgical removal of the testes and epididymis or a treatment which causes degeneration of the testes. The castrated, immature male, in addition to being sterile, fails to develop the secondary sex characteristics of the intact male. The accessory sex glands fail to develop, he does not develop the aggressiveness typical for the intact male of the species, and he does not develop sex drive. Castration of the mature male results in sterility, loss of sex drive in a short period of time, and the regression of the secondary sex glands. However, the secondary sex characteristics remain at the stage of development they had attained at the time of castration. Castration in horses is deliberately delayed until the secondary sex characteristics are developed for this reason.

Equipment commonly used for castration is shown in Figure 18-7. Sur-

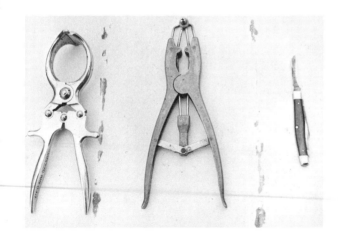

FIGURE 18-7. Equipment commonly used for castration. Left to right: emasculator, rubber band, applicator, and knife.

gical castration of lambs and calves is best accomplished by cutting away one cm of the bottom of the scrotum (Figure 18-8). This cuts through the tunica vaginalis exposing the testes. Each testis is then pushed through the opening, gripped with the thumb and forefinger, and removed with a quick jerk. When the blood vessels of the spermatic cord are broken in this manner, the stretching involved results in the lumen closing to prevent bleeding. When pigs are castrated an incision is made near the bottom of each scrotal pouch to expose the testes and the epididymis. The testis is separated from the tunica vaginalis and the spermatic cord is macerated rather than cut to prevent bleeding.

Nonsurgical procedures may be used for castrating lambs and calves. Emasculation is a procedure in which a special instrument called an emasculator is used to sever the spermatic cord without breaking the skin (Figure 18-8). This process disrupts blood and nerve supply to the testes resulting in degeneration. Rubber bands of a special design with an instrument for application are also available. The instrument (Figure 18-8) stretches the rubber band so that it can be slipped over the testes and released on the scrotum above the testes stopping the blood supply to the testes, again resulting in degeneration.

18-6.2
Vasectomy

Vasectomy is a surgical procedure in which the vas deferens are severed; usually a section is removed, resulting in sterility. Since the blood and nerve supply to the testes are not interrupted the male remains normal in all other respects. The vasectomized male has been used as an aid to estrus detection in the female when AI is being used in the herd or flock. The animal is effective in identifying the estrous female but there is a distinct disadvantage in that he can spread venereal diseases since he is able to copulate. For this reason penilectomy is sometimes done.

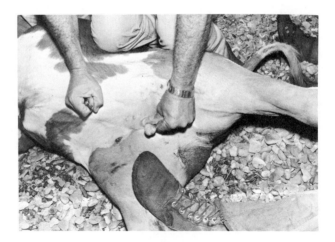

FIGURE 18-8. Three castration procedures: (a) knife is used to open scrotum so testes can be removed; (b) rubber band is placed around scrotum above testes; (c) emasculator is used to pinch spermatic cord.

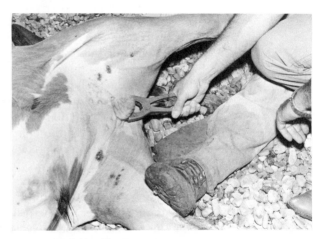

**18-6.3**
**Penile blocks**

The penile block is a simple procedure in which a plastic tube is inserted into the sheath. A trocar and canula are used to punch a hole through the sheath in line with an opening through the plastic tube (Figure 18-9). The stainless steel canula serves as a pin to hold the tube in place. The tube and pin prevent the protrusion of the penis and thus inhibit copulation. The procedure is effective for a relatively short time, but since the male is unable to copulate and ejaculate most tend to lose sex drive rather rapidly. Pain may be a factor in loss of libido. There is also some problem related to infection of the sheath where the puncture was made.

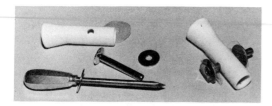

FIGURE 18-9.   Left: Unassembled penile block. Trocar is used to punch hole through sheath for application. Right: Assembled.

**18-6.4**
**Redirecting the prepuce**

This procedure has been performed on bulls but can also be done for rams. The purpose is to move the opening of the prepuce to one side so that the penis fails to line up with the vulva of the female, thus preventing copulation. It is accomplished by excising a small section of skin around the prepucial opening. An identical section of skin is removed 10 to 20 cm to one side. From this second opening a probe is used to loosen the skin from the abdominal wall so the sheath with the penis and the section of skin can be brought through to the new location and the skin sutured into place. The section of skin removed from the new location is used to cover the opening left at the old location (Figure 18-10). Males with redirected prepuce are

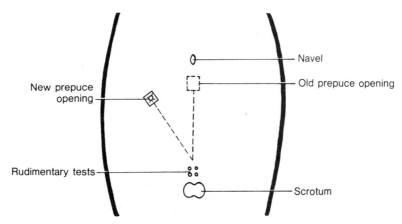

New prepuce opening

Navel

Old prepuce opening

Rudimentary tests

Scrotum

FIGURE 18-10.   Redirected prepuce procedure.

much more satisfactory for detecting estrous females than either the vasectomized males or males with penile blocks. Copulation is prevented, thus reducing the chance of spreading diseases. An additional advantage is that pressure of the male resting on the rump of the female allows the male to make an ejaculatory thrust and ejaculate. This prevents the male from losing sex drive, thus providing a useful tool over a much longer period of time.

**Suggested reading**

Beverly, J. R. 1978. Outlook for estrus synchronization. *The Advanced Animal Breeder*, p. 8. NAAB, Columbia, Mo.

Ericsson, R. J., C. N. Langevin and M. Nishimo. 1973. Isolation of fractions rich in human Y sperm. *Nature*, 246:421.

Foote, R. H. 1978. General principles and basic techniques involved in synchronization of estrus in cattle. *Proc. 7th Tech. Conf. Artif. Insem. Reprod.*, p. 74. NAAB, Columbia, Mo.

Hansel, W. and W. E. Beal. 1979. Ovulation control in cattle. *Beltsville Symposia in Agricultural Research. 3. Animal Reproduction*, p. 91. Allanheld, Osmun and Co. Publisher; Allsted Press, a division of John Wiley and Sons.

Hansel, W. and J. Fortune. 1978. The applications of ovulation control. *Control of Ovulation.* ed. D. B. Crighton, G. R. Foxworth, N. B. Haynes and G. E. Lamming. Buttersworth, London.

Lauderdale, J. W. and R. G. Zimbelman. 1974. Detection and synchronization of estrus. *Reproduction in Farm Animals.* (3rd.) ed. E. S. E. Hafez. Lea and Febiger.

Inskeep, E. K. and A. W. Lishman. 1979. Factors affecting postpartum anestrus in beef cattle. *Beltsville Symposia in Agricultural Research. 3. Animal Reproduction*, p. 277. Allenheld, Osmun and Co. Publisher; Allsted Press, a division of John Wiley and Sons.

Seidel, G. E. Jr. 1979. Application of embryo preservation and transfer. *Beltsville Symposia in Agricultural Research. 3. Animal Reproduction*, p. 195. Allenheld, Osmun and Co. Publisher; Allsted Press, a division of John Wiley and Sons.

Wiltbank, J. N. and E. Gonzalez-Padilla. 1975. Synchronization and induction of estrus in heifers with a progestrogen and an estrogen. *Ann. Bio. Anim. Bioch. Biophys.*, 15:255.

Wiltbank, J. N., J. C. Sturges, D. Wideman, D. G. LeFever and L. C. Faulkner. 1971. Control of estrus and ovulation using subcutaneous implants and estrogens in beef cattle. *J. Anim. Sci.*, 33:600.

Wishart, D. F. and S. B. Drew. 1977. A comparison between pregnancy rates of heifers inseminated once or twice after progestin treatment. *Vet. Rec.*, 101:230.

# Chapter

# 19  Reproductive management

Management plays an important role in the reproductive efficiency obtained from both females and males. Unfortunately reproductive efficiency approaching 100% is not possible even with the very best management; however, poor management can result in drastic decreases in reproductive efficiency.

**19-1**
**Measurements**
**of reproductive**
**efficiency**

In order to determine the effect of management on reproductive efficiency, it is necessary to establish some guide lines for measuring reproductive efficiency.

19-1.1
Services per
conception

This measurement is determined on a herd or flock basis by dividing total services by the number of pregnancies. Services per conception has little value for a large population of animals, but is a valid measurement for a single herd or an individual female. On a herd basis, unidentified sterile females will make the calculation less meaningful.

19-1.2
Calving rate

Calving rate is calculated by dividing the total number of cows bred by the number that calved. It is also expressed as percent calf crop. Lambing, farrowing, and foaling rates are calculated in the same manner. This measurement of reproductive efficiency is frequently inflated by culling known open females after the breeding season and using the remainder in the calculation. Such inflated values are meaningless in evaluating reproductive efficiency.

19-1.3
Nonreturn rates

With the advent of AI, a means of evaluating the fertility level of semen in the shortest possible time was recognized as a serious need. Pregnancy examination information was too scattered and not readily available. Actual calving rates were too expensive to obtain and came too late to be of greatest value. Artificial insemination organizations began calculating nonreturn (NR) rates at intervals following inseminations. A nonreturn rate is the percentage of females that do not return to estrus or receive a second service within a designated time interval. The time intervals most commonly used are 28–35 days, 60–90 days, and 150–180 days, Obviously, the shorter intervals provide earlier information but the longer intervals provide more accurate information. The difference between 28–35 day NR rate and 60–90 day NR rate ranges from 10 to 15 percentage units. The 150–180 day NR rate is only 1 to 2 percentage units lower than the 60–90 day NR rate. For this reason it is not often used.

Nonreturn rates are always higher than actual pregnancy rates because some open cows are not reinseminated. Some are bred naturally and some are culled or die without being reinseminated. Even though nonreturn rates are relative they do provide valuable information on males used in AI.

## 19-2
## Management related to the female

The main concern relative to when young females should be inseminated or bred for the first time should be size. The size at the time of breeding is important because it influences the size of the animal at first parturition. The significance of size at parturition relates both to uncomplicated parturitions and productivity of the female. Dystocia is a serious problem with undersized females. (See Table 22-2.) Larger females within a breed are more profitable producers.

## 19-2.1
## Size and age at first insemination

The effect of plane of nutrition on puberty and desired size at breeding is discussed in Chapter 22. The level of nutrition greatly affects the age at which puberty occurs, but once the female reaches puberty, neither size nor age affect conception rate within acceptable ranges of management. There are data indicating that heifers bred for the first time between 4 and 5 years of age experience a significant increase in reproductive problems. It is extremely important that females grow at a rate that will allow them to reach desired size compatible with the desired age at first breeding. The data in Table 19-1 show the desired weight at puberty, first breeding, and first parturition for the dairy breeds and some selected beef breeds. Heifers of the dairy breeds should reach the desired first breeding weight at an average of 15 months so that they can calve at approximately 24 months of age. Heifers calving at older ages through 36 months will produce slightly more milk during first and possibly second lactations than heifers calving earlier. Heifers that calve at 24 months on the average will have a higher lifetime production than those calving later.

Managers of many commerical beef cattle operations have followed a practice of having their heifers calve for the first time at 3 years of age. The major and perhaps only reason for this practice is to have the heifers large enough to avoid calving difficulties at first parturition. Managers need to take a look at the total cost of carrying the heifer to 3 years of age for first calving versus the cost of growing them fast enough so that they will be large enough to calve at 2 years of age. In addition to an earlier economic return, there is the added advantage of a shorter generation interval.

19-2.2
Detection of
estrus

In herds or flocks where artifical insemination is to be practiced, one of the most important management practices is detecting estrus so that insemination can be performed at the proper time. Estrus synchronization as described in Chapter 18 may relieve the problem somewhat especially with dairy heifers, beef cattle, sheep, and swine.

**19-2.2a Detection of estrus in cows** The problem related to estrus detection in cows is more critical than it is with other species because of their shorter and more variable length estrous periods. (See Chapter 5.) The data in Table 19-2 illustrate the need for frequent observation for detecting estrus. Using twice daily checks at 6:00 AM and 6:00 PM as a basis for comparison, an additional check at 12:00 noon increased efficiency by 10%. A fourth observation at 12:00 midnight gave an increase in efficiency of 19.9% over the twice daily checks. Estrus expectancy lists were used as an aid in the detection of estrus. Checking cows for estrus four times a day not only detects a higher percentage of cows in estrus so that they can be inseminated, but the beginning of estrus is more nearly determined so that insemi-

TABLE 19-1. Recommended body weight and size of heifers at puberty, first breeding, and first calving for the dairy breeds and selected beef breeds

| Breed | Puberty | 1st breeding | | 1st calving |
| | | Wt. | Heart girth | |
|---|---|---|---|---|
| | (kg) | (kg) | (cm) | (kg) |
| Ayshire | 232 | 284 | 152 | 432 |
| Brown Swiss | 272 | 340 | 160 | 500 |
| Guernsey | 215 | 272 | 150 | 410 |
| Holstein | 272 | 340 | 160 | 500 |
| Jersey | 170 | 250 | 147 | 385 |
| Angus | 260 | 280 | 151 | 400 |
| Hereford | 260 | 280 | 151 | 400 |
| Brahma | 300 | 320 | 148 | 420 |

TABLE 19-2. Cows detected in estrus when checked three or four times daily compared to twice daily.

| No. checks | Time checked | | | | % increase |
| | AM | Noon | PM | Midnight | |
|---|---|---|---|---|---|
| 2 | 6:00 | | 6:00 | | Base |
| 3 | 6:00 | 12:00 | 6:00 | | +10 |
| 4 | 6:00 | 12:00 | 6:00 | 12:00 | +19.9 |

From Hall, et al., J. Dairy Sci. 42:1086, 1959.

nation can be timed more accurately. The early morning check for estrus (as soon after daylight as possible) is the most important. More cows come into estrus between 2:00 AM and 5:00 AM than any other similar period of time during the day. (See Table 19-3.)

Checking cows for estrus should not be combined with other chores. It needs the full attention of the individual doing the checking. If feeding is done in connection with checking for estrus the cows will have their mind on eating rather than demonstrating symptoms of estrus. The individual observing cows for estrus should be thoroughly familiar with all of the physiological and psychological symptoms of estrus (Chapter 5). He should move among the animals to be checked causing as little distraction as possible. During the early morning check if the cows are still lying down they should be gotten up and moved around briefly and then carefully observed (see Figure 19-1).

TABLE 19-3.  Cows detected in estrus during AM versus PM

|  | Number | Percent |
| --- | --- | --- |
| AM | 32,405 | 72.5 |
| PM | 12,302 | 27.5 |
| Total cows | 44,707 | 100 |

Adapted from Foote, Search 8:1, Cornell University Agricultural Experiment Station, 1978.

FIGURE 19-1.  When checking cows for estrus, the one that stands when mounted is in estrus.

The livestock manager should use all aids available to him for detecting estrus. Section 18-6 describes procedures for altering males for use in detecting estrus. The penile block and redirected prepuce are recommended for use with cattle. A chin-ball-marker should be attached to the bull so that he will identify the estrous females. Dye preparations are available that can be painted onto the brisket for this purpose also. Cows treated with androgens have been used to help diagnose estrus in cows.

Some dairymen make a mark on the tail head with a grease-type cattle marker. When cows stand for mounting the mark becomes smeared. Heat check patches are available that can be glued to the tail head of the cow that will greatly increase the chances of detecting estrus. (See Figure 19-2.) The patch has a small capsule of red dye contained inside the larger white capsule. When the animal is in estrus and stands to be mounted, the pressure applied by the mounting animal will release the red dye causing the large capsule to turn bright red. A disadvantage of the heat check patch is that some cows will lose them or rub them off before coming into estrus. The number lost is not great and with daily observations the lost patches can be replaced with a minimum of effort.

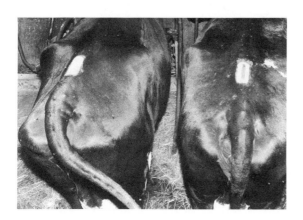

FIGURE 19-2. KaMaR mount heat detectors aid detection of estrus. The detector on right has turned red due to pressure from mounting cow.

**19-2.2b Detection of estrus in the ewe** Ewes do not demonstrate any signs of estrus when separated from the ram. Therefore, it is necessary to use an altered ram to detect ewes in estrus if AI is to be used. The vasectomized ram has been used most frequently, but rams with redirected prepuce are preferred because of reduced possibility of disease transmission. A dye material painted on the brisket or harness of the ram is usually used to identify the estrous ewe (Figure 19-3). The ewe flock should be checked for estrus at least twice daily.

**19-2.2c Detection of estrus in the sow** Sows usually demonstrate sufficient symptoms of estrus to be detected. The vulva of the estrous sow swells noticeably and the increased vascularity causes a reddish color particularly in the white breeds. There will be no visible mucus discharge from the vulva. Sows demonstrate the symptoms of estrus more plainly when boars are kept within hearing and smelling range. Both sows and boars produce pheromones that can be detected by the sense of smell. Sows respond to pheromone odor as well as noise made by the boar. Sows can be checked to determine whether they are in standing estrus by applying pressure to the lumbar region. If she is in standing estrus, she will assume the mating stance (Figure 19-4).

FIGURE 19-3. The ewe on the left has green chalk on rump from ram's harness. Technique can be used to record natural breeding dates or to detect ewes in estrus for AI.

FIGURE 19-4. This sow has assumed the mating stance from pressure applied by attendant's hand. Note swollen vulva.

**19-2.2d Detection of estrus in the mare**  Usual symptoms of estrus in mares are frequent urination, followed by a series of contractions of the vulva which expose an erected clitoris—a process referred to as "winking" (see Figure 19-5). The winking may continue for 2 to 3 minutes. Because of the variability in the behavior of mares, the most accurate means of detecting estrus involves teasing the mare with the stallion. If the mare is in estrus she will permit the stallion to bite her, will lower her rump and lift her tail to one side in addition to frequent urination and winking. If she switches her tail and attempts to fight the stallion she probably is not in estrus. Frequent urination and winking may occur during winter anestrus or during proestrus. The horse breeder should know the behavior of his mares especially during estrus. Some mares show strong signs of estrus, while a few never accept the stallion. In the latter case the breeder should palpate the ovaries to follow follicle growth so that the mare can be bred artificially when a follicle reaches 35 mm.

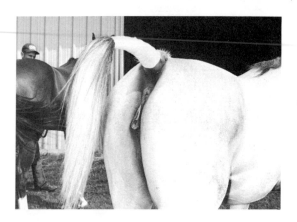

FIGURE 19-5.   Estrous mare demonstrating the "winking" phenomenon following urination.

**19-2.3
Timing of
insemination**

Proper timing of artificial insemination or hand mating is essential for optimum conception rate. The length of sperm life in the female reproductive tract, sperm capacitation, and viable life of the ovum have been discussed in previous chapters, but all relate to the optimum timing of insemination. For most species sperm life in the female tract is considered to be about 24 hours. It is reported to be longer in horses. Sperm capacitation in the rabbit takes about 6 hours, but may require a longer time in other species. The ovum remains viable up to 8 to 12 hours, with a better chance of survival when a capacitated sperm comes in contact with it soon after ovulation. The length of the estrous cycle and the time of ovulation in relation to the beginning of estrus also play an important role in the timing of insemination. Working within these confines recommendations will be given for each of the four species. (Further discussion Chapter 6.)

**19-2.3a Cows**   Figure 19-6 shows the conception rate from inseminations performed at intervals from the beginning of estrus until after ovulation. It should be noted that the best conception rate is obtained from approximately the middle of estrus until about the end of estrus. The conception rate for those inseminations performed near the beginning of estrus is considerably lower indicating that in many cases sperm have lost their

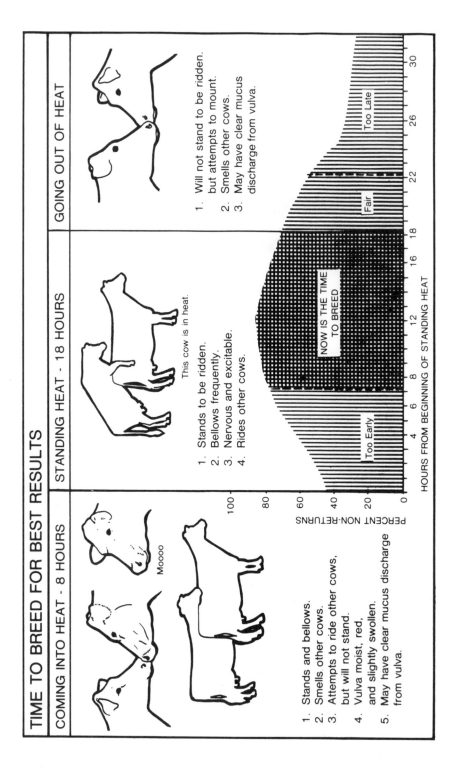

FIGURE 19-6. Expected average conception rates at intervals following the beginning of estrus. (From Asdell and Bearden. *Cornell Ext. Bull. 737.* 1959.)

viability prior to the time of ovulation. It also appears that inseminations occurring less than 6 hours prior to ovulation also result in lowered conception. Ova may have aged before capacitation in such cases. Some conceptions may result from inseminations performed after ovulation has occurred. Part or all of these may result from cows ovulating considerably later than expected.

A good rule of thumb to follow in a practical management situation is for cows first observed in estrus in the morning to be bred late the same day. Those cows first observed in the afternoon should be bred early the next morning. When cows are observed for estrus at mid-day those detected in estrus for the first time should be inseminated as late as possible the same day. (See Tables 19-4 and 19-5.) Accurate diagnosis of estrus is essential for proper timing of breeding. The data in Table 19-6 show the difference in nonreturn rate for cows in four management systems related to estrus detection. Cows turned out once and twice daily for estrus detection showed 5 and 6 percentage units higher nonreturn rate respectively when compared with those that remained in stanchions 24 hours a day. These data did not measure the difference in undetected estrus but based on the data in Table 19-2 one can assume a significant increase in percentage of detected estrus for those cows turned out twice daily.

TABLE 19-4.   Time of insemination and 150-180 day % NR

| Time observed in estrus | Time inseminated | Cows | Non-returns |
|---|---|---|---|
| | | no. | % |
| Morning (AM) | Before noon, same day | 1,308 | 67.1 |
| | Noon to 6 pm, same day | 27,320 | 69.9 |
| | After 6 pm, same day | 3,509 | 68.9 |
| | Before noon, next day | 268 | 62.7 |
| Evening (PM) | Before noon, next day | 6,893 | 69.9 |
| | Noon to 2 pm, next day | 4,948 | 67.4 |
| | After 2 pm, next day | 461 | 63.8 |

From Foote, Search 8:1, Cornell University Agricultural Experiment Station, 1978.

TABLE 19-5.   Recommended time for breeding cows in relation to onset of estrus

| Onset of estrus | Time to breed | Too Late to breed |
|---|---|---|
| AM | Same Day | Next Day |
| Noon | Late same day | Noon next day |
| PM | Early next day | Late next day |

TABLE 19-6.   Effect of winter turn-out management on nonreturn rate in cows

| Practice | 1st Services | No. Herds | % NR |
|---|---|---|---|
| No turn out | 9,412 | 494 | 64.1 |
| Out once daily | 47,365 | 3,237 | 69.5 |
| Out twice daily | 4,070 | 348 | 70.4 |
| Pen stable | 1,084 | 58 | 68.3 |

From Bearden, N.Y. Artificial Breeders Co-operator 132, 1956.

**19-2.3b Ewes**   Scientists still disagree as to when ewes should be inseminated for best conception. There seems to be more support for inseminating 12 to 18 hours after the onset of estrus than at other times. Most research has shown that two inseminations 24 hours apart will increase both conception rate and twinning rate. However, many do not feel that this increase is worth the extra time and effort required.

**19-2.3c Sows**   The preferred time for breeding the sow is approximately 24 hours after the onset of estrus. The estrus period in the sow lasts for 2 and 3 days with ovulation occurring at approximately midestrus or a little later.

**19-2.3d Mares**   The optimum time for inseminating or hand breeding the mare is more difficult to determine than for the other species. Ovulation occurs near the end of estrus whether the mare has a 3-day or 8-day estrous period. Best results without palpation are obtained by multiple breedings starting on the third day and repeating at 48-hour intervals until the mare is no longer in estrus. When only one breeding is desired because of a heavy breeding schedule for the stallion it is recommended that the mare be palpated and bred when she has a 35-mm follicle. She should be palpated 2 days later to see if ovulation occurred and if not, she should be rebred. When two large follicles are detected by palpation, mares should not be bred since pregnancies involving twins are usually terminated by abortion. Some breeders inject LH at the time of breeding to insure ovulation while sperm are viable.

**19-2.4
Rest after
parturition**

Sexual rest following parturition is more of a problem with cows than with other species. As described in Chapter 9, uterine involution in the cow includes the return of the uterus to the pelvic area, return to its nonpregnant size, and a recovery of normal uterine tone. The average time required for this is 45 days (Figure 19-7). However, histological studies have shown that another 15 days may be required before the endometrium is histologically normal. Dairy herd managers should palpate the involuting uterus at 35 days

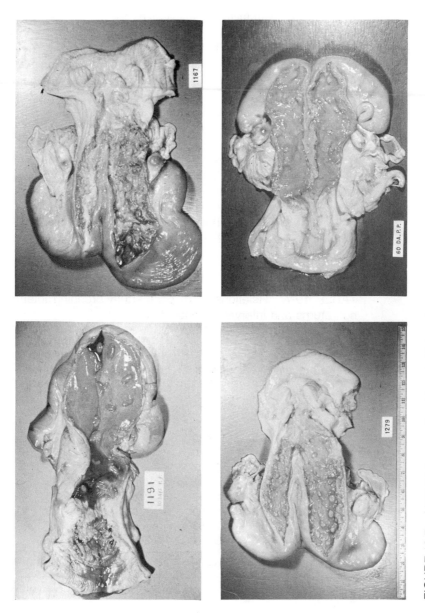

FIGURE 19-7. Uteri from cows slaughtered at intervals postpartum. (a) No. 1191, 10 days; (b) No. 1167, 20 days; (c) No. 1279, 46 days; and (d) 60 days. (Courtesy of N. L. VanDemark and Univ. of Ill., Urbana, Ill.)

postpartum to determine whether normal progress is being made. Treatment should be provided for those cows not making normal progress.

Based on this information researchers have for many years recommended that cows not be bred until the first estrus occurring after 60 days postpartum. The conception rate at this time will be 10 to 15 percentage units higher than those cows bred one estrous period earlier. (See Table 19-7.) Following this recommendation about 90% of the cows in a well-managed herd will maintain an average calving interval of approximately 12 months. The remaining 10% are classified as problem cows and even with veterinary assistance up to 5% of these cows may never conceive.

During recent years, a few individuals have recommended that cows be bred at the first estrus after 40 days postpartum. This recommendation has been made for herds with average calving intervals of 14 to 15 months as a means of shortening the calving interval as near to 12 months as possible. To the authors this is an attempt at finding a solution while avoiding the real problem, which in most cases is poor management. We do not think that the introduction of a questionable management practice to camouflage other poor management is acceptable. Perhaps 45 to 50% of the cows bred between 40 and 60 days postpartum will conceive to first service and these cows will have an average calving interval of only 330 days (50 plus 280). In high producing herds, all calving intervals as short as 330 days should be avoided for optimum milk production and certainly there is no justification

TABLE 19-7.  Relationship of postpartum breeding interval to percent nonreturns and interval to conception

| Interval, calving to 1st breeding | Cows | 150-180-day nonreturns | Interval, calving to conception | Projected calving interval |
|---|---|---|---|---|
| days | no. | % | days | days |
| 0-19 | 47 | 40 | 64 | 342 |
| 20-29 | 208 | 46 | 58 | 336 |
| 30-39 | 459 | 45 | 64 | 342 |
| 40-49 | 1,170 | 51 | 70 | 348 |
| 50-59 | 2,203 | 59 | 76 | 354 |
| 60-69 | 3,192 | 65 | 85 | 363 |
| 70-79 | 3,412 | 66 | 93 | 371 |
| 80-89 | 2,899 | 69 | 103 | 381 |
| 90-99 | 2,059 | 69 | 113 | 391 |
| 100 | 3,984 | 71 | 156 | 434 |
| All | 19,633 | 66 | 102 | 380 |

Adapted from Foote, Search 8:1, Cornell University Agricultural Experiment Station, 1978.

for having 40% of a herd with calving intervals that short. Furthermore, the same cows that become pregnant to the service between 40 and 60 days would also conceive to first service between 60 and 80 days and in the latter case, would have acceptable calving intervals.

Short postpartum intervals usually are not a problem with beef cattle. The beef cattle manager usually is more concerned with getting his cows started cycling so that they can be serviced early enough to maintain 12-month calving intervals. Refer to section 18-2 for treatment of anestrus.

Mares will come into estrus from 7 to 12 days after parturition (foaling heat) and can frequently be rebred with good results. However, mares should be bred at this time only if they have been given a careful examination to determine if there has been adequate recovery since parturition. A good rule of thumb is to check the mare on day 7, and if there is any question about her recovery, wait until the next estrus, which will occur about 30 days postpartum. Reasons not to breed at foaling heat include: (1) bruised cervix; (2) lacerations of the cervix or vagina; (3) vaginal discharge; (4) lack of uterine tone; or (5) the placenta was retained more than 2 hours.

Sows have a farrowing heat a few days after parturition. Fertilization is low at this time because of failure to ovulate. When they are rebred at the next estrus which follows weaning, uterine recovery will be complete and normal ovulation will occur. Some managers prefer to breed sows at the second estrus following weaning of pigs.

**19-2.5
Reproductive
records**

A complete set of records is absolutely essential for good reproductive management. Records which should be kept on each female are: (1) permanent identification; (2) parturition date; (3) date of first estrous after parturition and all subsequent estrous dates; (4) breeding dates with identification of service sire; (5) results of preliminary pregnancy checks between 35 and 40 days post-breeding; (6) results of final pregnancy check at 60 days or as soon thereafter as possible; (7) calculated due date; (8) date to turn dry (dairy); (9) actual parturition date; and finally (10) identification, sex and disposition of offspring. In addition, notation should be made at parturition of abnormal occurrences such as dystocia and retained placenta. Any abnormal discharge including cloudiness or flakes in estrous mucus should be noted. Record all treatments, whether performed by the livestock manager or by a veterinarian, along with a diagnosis and the date. A general herd health record is desirable and can be maintained along with the reproductive records. Include such items as diagnosis and treatment for mastitis, all vaccinations, and the diagnosis and treatments performed for any other health problem.

We recommend that a daily barn sheet be filled out (Figure 19-8). A record station should be established at a convenient location so that daily events may be recorded as they occur. A pocket notebook and pencil should be carried so that notes may be made while going about daily routines. These notes will help make the daily barn sheet record more accurate and complete. The manager should check the daily barn sheet at the end of the day to

DAILY BARN REPORT

Date  **8-4-79**

1.  Cows in milk **97** Dry Cows **37** Bred heifers **38** Carry-over heifers **87**
    Heifer calves **3** Bull calves **4**
2.  List cows giving sex of calf, tattoo, and chain no.

    _____

    _____

    _____

3.  Cows turned dry:_____

4.  Animals in heat: ___**458, 541, 605, 592**_____

5.  Cows and heifers bred.  List service bull.
    ___**541 - Victory**_____

    _____

6.  Milk sold_____ Milk fed calves _____ Milk produced_____

7.  Herd pasture **WP - Millet, Amsoo**Heifers on pasture_____
    Silage fed:  Milk cows **—** Dry cows **—** Heifers **—**
    Hay fed:  Milk cows **—** Heifers **—** Calves **—**
    Grain fed:  Milk cows_____ Dry_____Heifer_____ Calves_____

8.  Animals not bred--Expected in heat_____

9.  Animals bred--due to recycle if not pregnant_____

10. List animals treated giving type of and reason for treatment:
    ___**468 - Edema, LASIX ORAL**_____
    ___**676**   "         "        "_____
    ___**468**   **Mastitis - RR, LR-CEPHALAK  IM Temp 103°**___
    ___**490**   **Footrot - PENSTREP IM**_____

    _____

11. List animals died giving causes:_____

    _____

12. List animals sold, price, purchaser and address:
    ___**Dr. Trammell wants 2 bull calves**_____

    _____

    _____

    _____

FIGURE 19-8.   A daily barn sheet used in a dairy herd.

insure that all information has been recorded. The daily records are temporary and should be transferred to a permanent record system.

 The livestock manager needs to organize his records so that the reproductive status of the animal can be monitored daily, simply, and easily. For
dairy herds a good management tool, the *Dairy Herd Monitor,* is available
(Figure 19-9). It is designed as a revolving circle, divided into the 12 months
with the respective number of days for each month. Color coded, self-

adhesive removable stickers are used to identify each cow's breeding status. Blue represents fresh and open cows. Red represents bred cows and cows programmed for breeding. Previous estrous dates remind the manager when to expect cows in estrus. First and repeat breedings are registered by recording service sire's name or number on sticker. Green represents confirmed pregnancy. Yellow represents dry cows. The manager is reminded to dry treat for mastitis and it gives a countdown of days to parturition. The revolving circle is rotated one space daily to correct calendar date. With this tool no cow is lost with respect to reproductive status. The herd manager is constantly reminded of what needs to be done, when it needs to be done and what to expect and when.

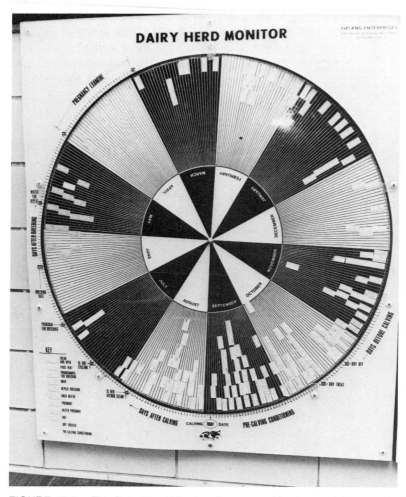

FIGURE 19-9.   The Dairy Herd Monitor gives a graphic picture of the reproductive status of a herd.

Data from the daily barn sheet should be transferred to permanent records on a regular basis. Permanent records include:

**1. Individual female card.** The information pertinent to each female should be transferred from the daily barn sheet to the individual female card. This card provides a lifetime breeding and health history (Figure 19-10).

**2. Breeding book.** This should be a hardback ledger book. Each breeding should be recorded in chronological order. The date, female identification number, service number and service sire identification should be recorded for each breeding. Not only is this information valuable from a reproductive point of view, but is also essential in providing the necessary information for registering offspring in a purebred herd (Figure 19-11).

**3. Offspring book.** This should also be a hardback ledger book. A sample page from an offspring book is shown in Figure 19-12. On the left hand sheet, tattoo number, date of birth of offspring, sex, and dam and sire identification are recorded. The right hand page is used for ear tag number, immunization and worming records, and disposal records.

HERD NO. 631
80
GP 83 (5-13-74)

| | JAN. | FEB. | MAR. | APRIL | MAY | JUNE | JULY | AUG. | SEPT. | OCT. | NOV. | DEC. |
|---|---|---|---|---|---|---|---|---|---|---|---|---|

NAME: MAGNOLIA ARCHIES GEM   REG. No. 7515888   DATE OF BIRTH 8-2-69   BIRTH WEIGHT   CONDITION AT BIRTH

SIRE: EMPEROR DUKE ARCHIE   REG. No. 1320310   TATTOO OR EAR TAG   RIGHT E589   LEFT E589   CALFHOOD VACCINATION   NO.   DATE   RUMEN MAGNETS

DAM: MAGNOLIA ALERT GEM   REG. No. 6292470   DEHORNED   EXTRA TEATS REMOVED   DATE & REASON FOR REMOVAL FROM HERD

L593

| MASTITIS & UDDER HEALTH | | | | | BREEDING RECORD | | | | | | CALF RECORD | | |
|---|---|---|---|---|---|---|---|---|---|---|---|---|---|
| DATE | RF LF LR RR | TREATMENT | HEAT DATES | DATE BRED | SERVICE BULL | PREG. CHECK | DATE DUE | DATE FRESH | SEX | NAME OR NO. | Sire & CALF DISPOSAL | | |
| 3/29/74 - | | 20cc Forte II in Udder | 12/2/71 | 12/15/70 | CBS Aetraneut | | | 11/6/71 | Dead | | Died | | |
| 3/30/74 - | | 20cc Forte II in Udder Penstrep IM | 12/25/71 | 1/27/71 | ABS Seaman | | | 12/12/72 | F | F116-505 | Bootmaker | | |
| 3/31/74 - | | 10cc Forte II & 10cc | 1/25/72 | 3/1/71 | ABS Bootmaker | | | 3/7/74 | F | F301-536 | Willis | | |
| | | | 4/14/74 | 2/17/73 | SS Elevation | | | 8/11/75 | F | F508-568 | Go | | |

**DIAGNOSIS AND TREATMENT BY VETERINARIAN**

REPRODUCTIVE DISEASES

| DATE | CONDITION | TREATMENT |
|---|---|---|
| 12/1/71 | Uterine Infection | 150cc Furacin, Penstrep, & Estrodiol |
| 12/14/72 | Fresh | Neomycin Bols in uterus |
| 1/25/73 | Estrus | 40cc Furacin in uterus for Experiment |
| 11/2/74 | Post Breeding Ut. Care | 40cc Penstrep, 120cc, Furocin, I.U. |
| 8/22/75 | Uterine Inj. | Penstrep & Furacin Intra V. |

OTHER DISEASES

| DATE | CONDITION | TREATMENT |
|---|---|---|
| 8/5/74 | Severe Mastitis (LF) | Sulfa Boluses orally, Penstrep IM |
| 8/6/74 | " " | 30cc Penstrep IM |
| 8/9/74 | " " | Sulfa Boluses orally, Penstrep, Predef 2X, Biosol IM |
| 8/10,11, 12/74 | " " | Penstrep, Predef 2x, Biosol, IM |
| 8/21/74 | " " | Penstrep, Predef 2x, Biosol, IM |

FIGURE 19-10. Lifetime record card used for individual animal.

312

| COW NO. | BREEDING DATE | SERVICE NO. | PREV. BREED. DATE | NEXT BREED DATE | SERVICE SIRE | COMMENTS |
|---|---|---|---|---|---|---|
| '43 | 8-1-79 | 4 | 7-10-79 | 8-21-79 | B, SOLDIER | REBRED |
| 301 | 8-2-79 | 1 | | 8-21-79 | CHOICE | REBRED |
| 323 | 8-2-79 | 3 | 6-23-79 | | MAVERICK | |
| 411 | 8-3-79 | 1 | | | CHOICE | |
| 541 | 8-4-79 | 2 | 7-8-79 | | VICTORY | |
| 394 | 8-8-79 | 5 | 7-20-79 | | MAVERICK | |
| 552 | 8-9-79 | 3 | 6-24-79 | | TIPPY | |
| 474 | 8-10-79 | 7 | 7-14-79 | | WHM BULL | |
| 305 | 8-14-79 | 1 | | | MAVERICK | |
| 203 | 8-15-79 | 1 | | | B.SOLDIER | |
| 690 | 8-16-79 | 4 | 7-24-79 | | VICTORY | |
| 347 | 8-16-79 | 1 | | | CHOICE | |
| 143 | 8-21-79 | 5 | 8-1-79 | | ASM | |
| 301 | 8-21-79 | 2 | 8-2-79 | | 7-G-178 (PGA) | |

FIGURE 19-11. Sample page from a breeding ledger book.

| TATTOO | DATE of BIRTH | SEX | DAM NO. | SIRE | EAR TAG NO. | WORMING AND VACCINATION | | | DISPOSAL RECORD |
|---|---|---|---|---|---|---|---|---|---|
| F760 | 11-20-76 | M | 9 | PRIDE | MALE | VAC. BRUCELLOSIS 3-9-77 | VAC. IBR, PI, (0NSAD) 3-10-77 | VAC. LEPTO-5- 4-9-77 | Sold 12/2 Auction |
| F761 | 11-20-76 | M | 429 | HOLLIREX | MALE | | WORMED-TRAMISOL 3-10-77 | VAC. BLACKLEG 5-10-77 | DEFORMED-DIED |
| F762 | 11-22-76 | M | 37 | ROYAL B | MALE | | | | Sold 12/2 Auction |
| F763 | 11-24-76 | F | 17 | PRIDE | 86 | | | | |
| F764 | 11-26-76 | F | 435 | CHOICE | 319 | | | BKD. | Sold 12/2 Auction |
| F765 | 11-29-76 | M | 470 | CHOICE | MALE | | | | DIED 12/2 SCOURS |
| F766 | 12-04-76 | F | 160 | MECURY | 214 | | | | |
| F767 | 12-09-76 | F | 20 | BOY | 89 | | | | |
| F768 | 12-09-76 | M | 347 | TOP HORNET | MALE | | | | Sold 12/16 Auction |
| F769 | 12-10-76 | M | 725 | BRUCE | MALE | | | | Sold V6 Auction |
| F770 | 12-10-76 | M | 611 | APOSTLE | MALE | | | | Sold 12/16 Auction |

FIGURE 19-12. Sample pages from offspring ledger book. Note vaccination and worming record on right.

The Dairy Herd Improvement Association (DHIA) record program is available to dairy herd managers. This is an official computerized record keeping program sponsored by the state agricultural colleges. In addition to production records certain information pertaining to the reproductive status of the individual cow and the entire herd is provided. Information is recorded on a monthly barn sheet which is mailed to a regional dairy record processing center. A computer printout is mailed back to the dairyman in about 10 days. The reproductive status of the individual cow is updated monthly and shows the number of days open or number of days pregnant along with the last breeding date. A reproductive efficiency summary is provided for the herd. The number of cows open less than 60 days, the number open for 60 to 100 days, and the number open more than 100 days is shown. Total number of pregnant cows, average days open, projected minimum calving interval, number of breedings during the past 12 months, conception rate, and percentage of problem cows are also provided.

There are some other computerized record systems. Some are sponsored by the Cooperative Extension Service and some are commercial. These systems usually provide production and management information and a complete set of financial records.

19-2.6
Detecting
reproductive
problems

Records mentioned in the preceding section are absolutely essential for early detection of reproductive problems. Early detection is necessary for two reasons: (1) problems are much easier to correct when detected early; and (2) problems corrected early result in less loss of reproductive time and money.

Before one can detect problems it is necessary to have standards for comparison. The following criteria can be used for this purpose in dairy herds, and can serve as guidelines for beef cattle and other species.

1.   At least 90% of the cycling cows (after 40 days postpartum) should be detected each estrous period. This will require at least two and preferably three checks for estrus per day.

2.   From 60 to 65% of the cows will calve to first service. Pregnancy diagnosis at 60 days postpartum will also reflect these percentages.

3.   There should be no more than 10% of the cows classified as problem cows at any one time. Problem cows are those that (1) fail to conceive to three services; (2) continue to have abnormal discharges and/or uninvoluted uteri beyond 60 days postpartum.

4.   The number of cows leaving the herd because of failure to breed should not exceed 5% in any 12-month period.

5.   The average number of services per pregnancy should be two or less. In a 100-cow herd this would allow for 100 first services to which 60 cows become pregnant, 40 second services with 20 becoming pregnant, and 20 third services with 10 becoming pregnant. This accounts for 90 cows with

160 total services and leaves up to 40 services for the remaining 10 problem cows. The actual number of services utilized will depend on the extent of the problems and how soon the nonbreeders are eliminated from the herd.

Managers should be especially alert for:

1. Abnormal discharge (blood or pus) at any time. A small amount of blood without pus about 24 to 36 hours after estrus is probably normal metestrus bleeding and should cause no alarm.

2. An open cow not observed in estrus by 50 days postparturm. The cow may be in anestrus, but if palpation reveals a corpus luteum on an ovary she is cycling but not being detected in estrus.

3. Estrous cycles of irregular length (less than 18 days or more than 24 days). Short cycles are an indication of ovarian dysfunction and long cycles are an indication of embryonic or fetal loss or undetected estrus.

4. Continuous or prolonged estrus (more than 24 hours). This condition is an indication of follicular cysts.

5. Retained placenta. From 5 to 15% retained placentae can be expected in healthy herds (Table 19-8). There seems to be a breed difference between dairy breeds as well as a lower incidence in beef cattle than with dairy cattle.

6. Observable abortions. Every cow that aborts should be suspected of having a contagious disease. A veterinarian should be called in to make a diagnosis in each case.

7. Cows that have not settled to three services. These cows are classified as problem cows and veterinary assistance may be desirable.

TABLE 19-8. Incidence of retained placenta in the Mississippi State University dairy herd by breed (Jan. 1971-Dec. 1977)

| Breed | Mean % |
|---|---|
| Ayrshire | 25.5 |
| Jersey | 2.3* |
| Guernsey | 28.2 |
| Holstein | 22.1 |
| Brown Swiss | 27.3 |

*Means differ (P<.05).
Watts, et al. Proceedings So. Div. ADSA, 1979.

In most herds problem situations are usually related to individual cows. Any one of the above mentioned symptoms is serious for that particular cow and she should be handled as an individual case. These symptoms do not necessarily indicate an infectious disease. However, disease should not be ruled out, as many of the listed symptoms can be caused by one or more infectious diseases. If more than 10% of the herd is affected by a combination of these symptoms, a herd problem probably exists. Whether the problems relate to individual cows or the entire herd it is essential to act quickly to correct the problems. Early treatment increases the likelihood of successful correction of the problem and reduces losses from delayed breeding and sterility.

Prevention is far superior to cure. A good herd health program carried out in conjunction with a veterinarian is by far the most economical approach to maintaining a healthy herd. In such a program the veterinarian should visit the herd at least once a month to provide the following services: (1) examine all cows between 30 and 60 days postpartum to determine if the uteri are involuting properly; (2) administer calfhood vaccination for brucellosis and perhaps other vaccinations depending on the expertise of the livestock manager; (3) examine and/or treat cows with mastitis problems; (4) examine and/or treat cows with reproductive problems; and (5) perform pregnancy diagnosis if livestock manager is not proficient in this practice. A herd health program of this nature should reduce the number of emergency visits by the veterinarian to treat actute problems.

## 19-3
## Male

The reproductive process in the male is probably just as complicated as it is in the female. However, the processes in the male are not influenced nearly so drastically by human management or environmental factors as they are in the female.

## 19-3.1
## Age and size factors in semen production

The age at puberty has been discussed in Chapter 6 and will not be covered here except to reemphasize that level of feeding can speed up or slow down the onset of puberty. Size is also affected by level of feeding and from the time of puberty until the mature age is reached it is difficult to separate age and size, particularly in well-managed herds. The data in Table 19-9 shows semen characteristics of a Holstein bull during the four years following puberty. The ejaculate volume and total sperm produced per year more than doubled between the first and fourth years. Sperm concentration increased between the first and second year but remained essentially constant for the remaining years. The size of the bull is not given but it is reasonable to assume that the bull continued to grow even through the fourth year.

A number of studies have shown a highly significant correlation of 0.90 between testes size and body weight. Even though the relationship between testes size and body weight diminishes after puberty, larger males of the same species will usually have larger testes than smaller males.

TABLE 19-9. Semen characteristics of a Holstein bull ejaculated three times per week four consecutive years following puberty

| Characteristic | Year | | | |
| --- | --- | --- | --- | --- |
| | 1st | 2nd | 3rd | 4th |
| No. ejaculates | 153 | 151 | 143 | 135 |
| Av. volume/ejaculate cc | 2.8 | 3.8 | 5.0 | 6.0 |
| Sperm concentration ($10^6$/cc) | 1089 | 1530 | 1513 | 1451 |
| Total sperm produced ($10^9$) | 523 | 898 | 1093 | 1197 |

From VanDemark et al., J. Dairy Sci. 39:1071, 1956.

The relationship between testes size and sperm output has a significant correlation of 0.80. It is apparent then that the larger the male the larger the testes will be and the greater the output of sperm. Most work of this nature has been done with bulls but it also relates to the other species.

**19-3.2 Bull handling techniques**

Bulls in general, and dairy bulls in particular, can be dangerous and should be handled accordingly. Dairy bulls should never be allowed to run in a pasture or paddock in which humans need to enter. The student is referred to section 14-1 for facilities needed for handling bulls.

The frequency of collection of semen for artificial insemination will depend a great deal on the need for semen from a particular bull. If the goal is maximum spermatozoa harvest, the bull may be collected as often as his libido will allow. Bulls have been collected daily for an extended period (300 days) with no decrease in semen quality or conception rate. The volume and total sperm collected per day decreased in some bulls to the point of not being practical for commercial processing. In such situations two ejaculates may be collected every other day to achieve similar sperm harvest with a greater number of sperm available each processing day. In natural mating it is desirable to match each bull with the number of cows that he can handle without exhaustion during the specified breeding season. These bulls may be ejaculating several times per day for several consecutive days and if there are too many cows in estrus at one time some may not be bred or the bull's sperm supply may become depleted before all the cows have come into estrus.

Proper nutrition is extremely important for both the young growing bull and the mature bull. Overfeeding and underfeeding adversely affect health and libido of males. Further discussion can be found in Chapter 22.

*Exercise* was once thought to be essential for maintaining semen quality and fertility levels. More closely controlled experiments have disproven this idea by showing that bulls maintained for 6 months or more in tie stalls or box stalls maintain conception rates equal to those that were force exercised for 30 minutes per day. This is not to imply that exercise is unnecessary, however. It is reasonable to expect that bulls that have an opportunity to exercise regularly will experience less lameness and will require less foot trimming. There may be other beneficial effects from exercise related to maintaining a healthy animal.

*Transporting* bulls from one location to another need not be a factor in maintaining fertility. Care should be taken to maintain comfortable conditions for the bull that is being transported and to prevent injury. Moving the bull from familiar quarters to unfamiliar quarters does not seem to have any affect on fertility.

## 19-3.3 Reproductive management of the stallion

Young stallions should be separated from fillies when they are weaned. They will not have reached puberty by weaning, and separation at this time is easy to incorporate into the management routine. As 2-year-olds they should be limited to four or five mares. The goal of the breeder is to give his 2-year-olds training and experience. Three-year-olds can be used three to four times weekly, permitting twenty to thirty mares to be scheduled to them during the breeding season. Older stallions can be used once per day for 5 to 6 days a week or twice a day 3 times per week. This will permit the scheduling of about forty mares to an older stallion during a breeding season. Frequency of use will usually be limited by libido rather than by semen quality.

If demand for a stallion is high his total services can be increased by a factor of about ten during a breeding season through use of AI. Approximately 5 billion sperm can be collected from a stallion in a day. Using the recommended 500 million sperm per insemination, ten inseminations per day would be possible. To gain maximum utilization of semen from an AI program, synchronization of mares should be scheduled to insure that enough mares are in estrus when the semen is collected. Twice as many mares can be bred if they are palpated and inseminated when the follicle is 35 mm as compared to breeding every other day without palpation. The palpation technique would apply to both AI and natural mating.

When evaluating a stallion for purchase or use in a breeding program several items should be on the evaluation check list. The first three items are best evaluated during the breeding season. (1) Semen quality tests should include concentration, motility and morphology. Semen should be collected and evaluated, followed by another collection and evaluation 60 to 90 minutes later. (2) Reaction time or libido of the stallion should be checked with an estrous mare. During the breeding season expected reaction time is about 2 minutes. During the winter his reaction time may be about 10 minutes. (3) The temperament of the stallion is important both as to ease in handling and whether he chews excessively on the mares during teasing. Other items to evaluate include (4) his past breeding record, (5) his health record, and (6) his conformation.

**19-3.4
Semen quality
and health
checks for
males used in
natural mating**

A high percentage calf crop in herds being mated naturally is dependent to a large degree on using males capable of producing semen that has a high fertility level. Unfortunately there is no laboratory test that will determine the fertility level of semen. (See Chapter 15.) The best that can be done is to collect semen from the male and subject it to certain quality tests. The quality tests normally performed are gross examination and motility. In some cases, concentration and morphology of the sperm are also determined. The student is referred to Chapter 15 for procedures for conducting these quality tests.

The following procedure for bulls can be adapted to the other species. Based on semen quality, bulls can be classified into three categories; *satisfactory, questionable,* or *cull.* A satisfactory sample of semen will have a good concentration as indicated by an opaque milky-white color. The sample should have 40% or greater progressive motility and if morphology is performed the sample should have less than 25% abnormal sperm. Semen that falls below the satisfactory standards on one or more of the three criteria would be classified as questionable. Samples that are clear or watery indicating few or no sperm would be classified as cull. Samples with good concentration but with an extremely high level of abnormal sperm or very low motility should also be classified cull.

Bulls producing questionable or cull quality semen may be rechecked one month later if the owner desires. Experience shows that only about one out of every thirty bulls classified as questionable or cull will improve enough that the semen can be classified as satisfactory. The bull needs to be valuable to justify the cost of a recheck.

The semen from young bulls should be checked but one should remember that beef bulls usually reach puberty later than dairy bulls. Most bulls of the European breeds can be checked for semen quality at about 14 months of age but other breeds should not be checked until they are about 18 months old. Ejaculates from young bulls will have a lower volume and usually a lower concentration than ejaculates from older bulls.

Semen quality checks should be made on bulls 2 months before the breeding season. This will allow time for retest should such be desirable and it will also allow time to find replacement bulls for those that need to be culled.

Transmissible diseases should always be of concern in a natural mating program. The easiest way to maintain a disease-free herd once this status is attained is to maintain a closed herd. This means introducing no breeding animals, either male or female, from outside the herd. This frequently is not possible so it is necessary to use safeguards. When buying bulls for breeding it is safer to buy prepuberal bulls and keep them isloated until health checks can be performed. If breeding age bulls must be purchased it is absolutely essential that they be tested for certain communicable diseases. They must be tested for brucellosis, vibriosis, leptospirosis, and trichomoniasis. In addition it is desirable to test for the three more common viruses, infectious bovine rhinotracheitis, bovine virus diarrhea (BVD), and parainfluenza–3. The bull should be tested prior to purchase and kept isolated from the rest of the herd for 30 to 60 days until a second test can be performed.

**Suggested reading**

Bearden, H. J. 1956. Improving your herd's conception rate. *N.Y. Artificial Breeder's Co-operator,* 13:2.

Foote, R. H. 1978. Reproductive performance and problems in New York dairy herds. *Search,* 8:1. Cornell University Agr. Exp. Sta.

Hall, J. G., C. Branton and E. J. Stone. 1959. Estrus, estrous cycles, ovulation time, time of service and fertility of dairy cattle in Louisiana. *J. Dairy Sci.,* 42:1086.

Kiddy, C. A. 1979. Estrus detection in dairy cattle. *Beltsville Symposia in Agricultural Research. 3. Animal Reproduction,* p. 77. Allanheld, Osmun and Co. Publisher; Allsted Press, a division of John Wiley and Sons.

Maddox, L. A., A. M. Sorensen, Jr. and U. D. Thompson. 1961. Testing bulls for fertility. *Texas Agr. Exp. Sta.–Texas Agr. Ext. Service Bull.* 924.

Morrow, D. A., S. J. Roberts and K. McEntee. 1969. A review of postpartum ovarian activity and involution of the uterus and cervix in cattle. *Cornell Vet.,* 59:134.

Pickett, B. W. and J. L. Voss. 1973. Reproductive management of the stallion. *Proc. 18th An. Conv. Am. Assoc. Equine Pract.,* p. 501. ed. F. J. Milne.

Salisbury, G. W., N. L. VanDemark and J. R. Lodge. 1978. *Physiology of Reproduction and Artificial Insemination of Cattle.* (2nd ed.) W. H. Freeman and Co.

Trimberger, G. W. and H. P. Davis. 1943. Conception rate in dairy cattle by artificial insemination at various stages of estrus. *Nebraska Agr. Exp. Sta. Res. Bull.* 129.

VanDemark, N. L., L. J. Boyd and F. N. Baker. 1956. Potential services of a bull frequently ejaculated for four consecutive years. *J. Dairy Sci.,* 39:1071.

Watts, T. L., J. W. Fuquay and C. J. Monlezun. 1979. Season and breed effects on placenta retention in a university dairy herd. *Proc. An. Mtg. So. Div. ADSA.*

# Chapter

# 20

# Pregnancy diagnosis

**20-1**
**Cow**

The economic value of an early diagnosis of pregnancy in the cow is quite clear. Whether one is dealing with dairy cattle or beef cattle, the ultimate goal is an average 12-month calving interval for the herd. Every management practice which contributes to attaining this goal is well worth consideration. Pregnancy diagnosis is one of these tools. Most cows that fail to conceive will return to estrus in approximately 21 days post-breeding. Due to a number of different causes a small percentage do not. This latter group, even though small, are the ones that need to be discovered by pregnancy diagnosis as early as possible. Regardless of how good a management program might be, not all cows that return to estrus are detected. These cows also need to be discovered early so that additional attention can be given to heat detection so that they can be rebred.

The ideal pregnancy test would be one that is inexpensive and highly accurate, that could be conducted at the farm, by farm personnel, utilizing milk, urine, or other easily obtained specimens, that would detect pregnancy at 17 to 19 days post-breeding. No such test is available today. The milk progesterone assay test shows that progress is being made. This test is conducted on milk samples taken 21 to 24 days post-breeding. It has to be conducted in a laboratory with highly sophisticated and expensive equipment and is not available to most livestock producers. Therefore, diagnosis of pregnancy by rectal palpation is the only practical means available to most cattlemen today.

Some authors have recommended that diagnosis of pregnancy by rectal palpation be done by practicing veterinarians. We disagree and feel strongly that pregnancy diagnosis by rectal palpation is a good management tool that all progressive livestock managers should utilize. We would like it clearly understood that all cattle operations should have a good preventive herd health program in which the veterinarian is an integral part. A good relationship between the cattleman and the veterinarian will usually identify the practices that should be carried out by the herdsman and those that should be relegated to the veterinarian. A wise veterinarian will work with his client to help him become more proficient with his management tools.

In order for pregnancy diagnosis to be of greatest benefit, it is necessary to set up a workable schedule and follow it rigidly. We recommend for dairy herds and beef herds using artificial insemination that some pregnancy palpations be scheduled each week. This is best accomplished when the same time of the same day each week is designated for this purpose. The next decision to be made is when should cows be palpated following breeding? In herds where AI is used the first palpation should be made between 35 and 42 days after insemination. Even though palpation at this stage requires greater expertise, the main interest should be to identify those cows that are not pregnant so that they can be observed more carefully during the next few days when they should be returning to estrus. Many cows that are open at this time probably were in estrus 21 days post-breeding but were not detected. All questionable cows at this time should be placed in a group to be observed carefully for estrus. All cows that have not returned to estrus by 60 days post-breeding should be palpated a final time. A few pregnancies (1–2%) at 30 to 35 days will result in embryonic mortality and can be picked up as open at the 60-day check. Very few pregnancies terminate after 60 days except as a result of a disease which causes abortion. Accurate records of each palpation should be maintained for each individual cow.

The management system used for individual beef cattle operations will dictate to some extent when pregnancy palpations are to be made. If the management tool is to be of value in retaining breeding females in a herd, it must be performed as described above. Frequently pregnancy palpation is used simply to determine which cows to cull so that open cows are not carried through the winter. This can be accomplished in connection with other management practices which require running cows through a chute.

**20-1.1a Structures to be palpated**   The *cervix* is chiefly a landmark serving as a guide for locating other structures. It is easily identified by palpation. The position and to some extent size gives an indication of the stage of pregnancy, but a diagnosis should never be based on the cervix alone.

Most of the diagnosis is based on the *uterus* and its contents. The size of the uterus influences its position in relation to the pelvis and should be noted. The thickness and tone of the uterine wall are important. The uterine wall becomes thinner as pregnancy progresses and is very resilient to touch compared with the uterus of the open cow. This is particularly important in differentiating between pregnancy and a condition which causes uterine enlargement without pregnancy. A dorsal bulge is detectable from 30 to 50 days in the area of the body of the uterus. The uterine wall is thinnest at this point and the pressure created by the contents of the uterus causes the bulging effect.

The *chorionic membrane* can be detected by gently grasping the uterine wall between the thumb and forefinger and lifting slightly. With some practice one can feel the membrane slip from between the thumb and finger. Thus, the term "slipping the membrane" has been used to describe this procedure. By 120 days the placentomes are large enough to palpate through the uterine wall. At first these will be about the diameter of a dime and become much larger toward the end of pregnancy.

The *contents of the uterus* are the most positive diagnostic structures to be palpated. Between 30 to 50 days the amnion is quite turgid and can usually be detected by palpating the uterine horns with the thumb and forefinger. This can be done by starting near the tip of the uterine horn and applying gentle pressure as the thumb and forefinger are moved back toward the cervix. After 50 days the amnion begins to soften, and by 60 days it is no longer detectable. After 60 days the fetus can be palpated except during a period from 170 to 230 days when it is too deep in the abdominal cavity to reach in large cows.

The *ovaries* can be palpated up to about 120 days. Structures on the ovary can help confirm either a positive or a negative diagnosis. Pregnancy is always accompanied by a corpus luteum. However, one must remember that a corpus luteum is not always accompanied by a pregnancy. The absence of a corpus luteum would confirm that the cow is not pregnant.

The *pulse of pregnancy* can be a help in confirming a diagnosis, particularly at certain stages of pregnancy. This pulse is felt in the middle uterine artery which branches from the internal iliac and transcends to the uterus via the broad ligament on the right side near the forward edge of the pelvis (Figure 20-1). It is suspended in a fold of this ligament which permits it to be readily picked up through the walls of the rectum and moved about freely.

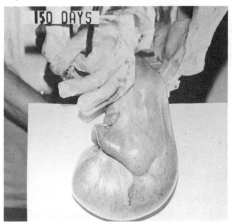

FIGURE 20-1. Palpation of a simulated middle uterine artery. Note finger completely surrounding the artery.

The uterine artery in the nonpregnant cow is 3 to 4 mm in diameter and has a clearly defined pulsation. In the pregnant cow, it begins to enlarge when the conceptus is large enough to require an increased blood supply and may reach a diameter of 1 to 1.5 cm by the end of gestation. The pulsation in the enlarged artery is much more forceful, indicating a greater volume of blood flow. By 120 days of pregnancy the middle uterine artery has enlarged sufficiently to be used as a differential diagnosis in pregnancy determination by rectal palpation.

**Palpation at 35 to 40 days.** This stage of pregnancy requires more skill than later stages. However, it can provide valuable management information when used properly. Figure 20-2 shows an open uterus on the floor of the pelvis for reference. The following features should be identified.

1. Uterus on the floor of pelvis, except in larger cows with elongated reproductive tracts. Slight enlargement of one horn with detectable dorsal bulging (Table 20-1). Thinning of the uterine wall with a fluid-filled feeling (Figure 20-3).
2. Slippage of the membrane possible.
3. An amnion about the size of the yolk of a hen egg can be detected.
4. Corpus luteum present on the ovary adjacent to the horn containing the amnion.

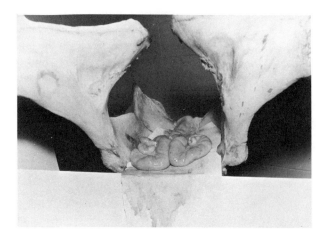

FIGURE 20-2. Nonpregnant uterus lying on the floor of the pelvic cavity.

**Palpation at 45 to 50 days.**

1. Uterus still on pelvic floor. Slightly greater difference in size of pregnant and nonpregnant horn, with the pregnant horn being 5 to 6.5 cm in diameter and the dorsal bulging more pronounced.
2. The amnion will be about the size of a small hen egg.

TABLE 20-1. Diameter of pregnant bovine uterine horn at different stages of pregnancy

| Stage in Days | Diameter in cm |
|---|---|
| 30 | Slight enlargement and dorsal bulging |
| 60 | 7 |
| 90 | 8 |
| 120 | 12 |
| 150 | 18 |

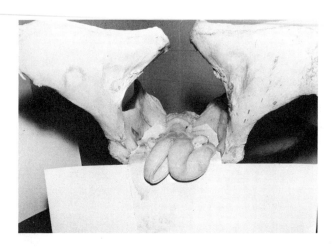

FIGURE 20-3. Pregnant uterus—35 days. Note bulge in left horn and CL on left ovary.

3. The membranes can be slipped in either horn.
4. The corpus luteum will be on the ovary adjacent to the pregnant horn.

**Palpation at 60 days.**

1. The pregnant horn will be dropping slightly over the brim of the pelvis and feels like a balloon filled with water (Figure 20-4). The pregnant horn will be 6.5 to 7.6 cm in diameter and the dorsal bulge is no longer detectable.
2. The membranes can be slipped in both horns but the amnion will not be detectable.
3. The fetus (5 cm long) can be bumped by rubbing the hand over the outer curvature of the uterine horn or by pressing against the outer curvature and then moving the hand slightly but quickly posteriorly (Table 20-2).
4. A corpus luteum will be on the ovary adjacent to the pregnant horn.

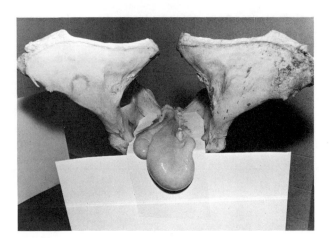

FIGURE 20-4. Pregnant uterus—60 days. Left horn containing fetus is pulled over brim of pelvis.

TABLE 20-2. Average size of bovine conceptus at different ages: variation occurs between breeds and within breeds

| Age/Days | Crown/Rump Length in cm |
|:---:|:---:|
| 30 | 1 |
| 60 | 5 |
| 90 | 13 |
| 120 | 30 |
| 150 | 38 |
| 180 | 56 |
| 210 | 71 |
| 245 | 81 |
| 280 | 86 |

**Palpation at 90 days.**

1. The uterus will be pulled well over the pelvic brim and will be 8 to 10 cm in diameter (Figure 20-5).
2. The fetus will be 10 to 15 cm long and will be easily palpated.
3. A corpus luteum will be on the ovary adjacent to the pregnant horn.

**Palpation at 120 days.**

1. The uterus will be well over the brim of the pelvis with the cervix pulled almost to the pelvic brim (Figure 20-6).
2. The fetus can be easily palpated and will be from 25 to 30 cm long. Anatomical parts of the fetus can be identified.

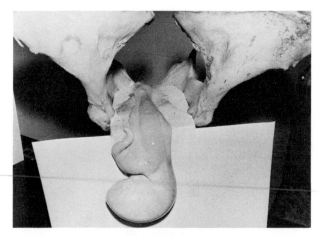

FIGURE 20-5. Pregnant uterus—90 days. Both horns are pulled over pelvic brim.

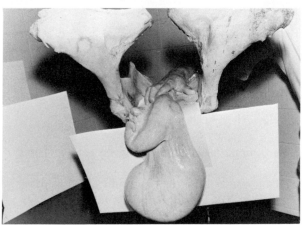

FIGURE 20-6. Pregnant uterus—120 days. Cervix is pulled to brim of pelvis.

3. Small placentomes can be identified.
4. The pulse of pregnancy can be detected.
5. The ovaries may be difficult to reach but a corpus luteum will be present on the ovary adjacent to the pregnant horn.

**Palpation at 150 days.**

1. The uterus will be pulled well into the abdominal cavity and the cervix will be located at the brim of the pelvis (Figure 20-7).
2. Distinct placentomes about the size of ovaries can be identified.
3. The fetus is well formed and will be 35 to 40 cm in length but may be difficult to reach in larger cows.
4. The pulse of pregnancy will be quite distinct with the artery being 6 mm to 1.25 cm in diameter.

FIGURE 20-7.  Pregnant uterus—150 days. Cervix may be pulled partially over brim. Uterus will be difficult to palpate.

### Palpation at 170 to 230 days.

1. Cervix will be at the brim of the pelvis and may be bent over the edge.
2. The dorsal wall of the uterus will be tight and difficult to palpate.
3. The placentomes will vary in size and may be difficult to palpate because of the tight uterine wall.
4. The fetus will be difficult to palpate particularly in larger cows due to the depth of the abdominal cavity.
5. There will be a strong pulse of pregnancy and the artery will be 1.25 to 1.4 cm in diameter.

### Palpation at 230 to 280 days.

1. The fetus will be large enough to extend back within range of the hand. The head and front feet are usually the structures palpated.
2. Movement of the fetus can fequently be detected (Figure 20-8).

**20-1.1b Differential diagnosis**  Pregnancy is by far the greater cause of uterine enlargement, but by no means is it the only cause. At each of the stages described, pregnancy must be differentiated from one or more of the other causes of uterine enlargement.

FIGURE 20-8. A squeeze chute with head gate should be available for palpating beef cows.

**1. Pyometra.** Pyometra is a condition characterized by an accumulation of pus in a sealed uterus. The condition occurs when an infectious organism enters the uterus at the time of or prior to the onset of pregnancy. The organism allows pregnancy to be initiated but after a variable period of time causes death of the embryo or early fetus. The conceptus is not expelled and degenerates in the sealed uterus accompanied by the formation of pus. The amount of pus may vary from a relatively small amount to several liters. In cases of this sort, the contents of the uterus prevent the release of $PGF_2\alpha$, which in turn results in the corpus luteum remaining functional. The condition may remain *status quo* for an extended period of time.

Pyometra differs from pregnancy in that the uterine wall is thicker, spongy, and less resilient. In addition, the pus is more viscous than the fluid of pregnancy and frequently can be moved from one horn to the other. Of course, there is no fetus to palpate. The stages of pregnancy which need to be differentiated from pyometra are 45 days through 120 days.

**2. Endometritis.** Endometritis is a nonspecific infection of the endometrium. It is characterized by the absence of pus and a uterine wall which feels thickened and spongy. This condition may be confused with pregnancy at the 35 to 40 day stage.

**3. Metritis.** Metritis is a nonspecific infection of the uterus characterized by the presence of visible pus. The pus may be seen on the lips of the vulva and on the tail where it rubs across the vulva. The pus may also be seen in estrous mucus as cloudiness or yellow or white flakes. The uterine wall is thickened and spongy to feel. This condition might be confused with the 35 to 40 day stage of pregnancy.

**4. Mummified fetus.** Mummified fetus is a condition in which the fetus dies and the fluids and soft tissues are reabsorbed (Figure 20-9). Depending on the stage at which the condition is detected, the mass may range from a semisolid to a solid ball. It is not difficult to differentiate between a mummified fetus and normal pregnancy at the 90 to 120 day period. However, with a casual palpation one might feel the mummified fetus and misinterpret it for a normal fetus. Additional palpation would reveal the absence of fluid surrounding the mass and provide the differential diagnosis.

FIGURE 20-9. A mummified fetus. (From Hafez. 1974. *Reproduction in Farm Animals.* (3rd ed.) Lea and Febiger.)

**20-1.2
Assay of
progesterone in
milk**

Researchers in Great Britain, Germany and the United States have reported pregnancy testing on the basis of progesterone levels in milk. Milk progesterone levels have been found to parallel blood progesterone levels when samples were taken periodically throughout the estrous cycle. The levels correspond to a pattern set by the corpus luteum. That is, high during midcycle and low just before, during and just after estrus.

The prediction of pregnancy using this test is based on the progesterone level in a single sample of milk taken at 21 to 24 days post-breeding. For cows producing positive samples (>11ng/cc) 80% have been pregnant 40 to 60 days post-breeding as determined by rectal palpation. Essentially all the cows producing milk negative to the test (<2 ng/cc) have been open.

The difference between the predicted pregnancy based on progesterone level at 21 to 24 days compared to actual pregnancy at 40 to 60 days post-breeding have been referred to as false positives. In reality a high percentage of those cows may actually have been pregnant at, or shortly before, the 21-to-24-day sampling period. Embryonic mortality then would be the reason for their not being pregnant when palpated 40 to 60 days post-breeding.

Research has shown that with virgin heifers free of known reproductive diseases, embryonic mortality accounts for 46 to 75% of the pregnancy failures when the heifers were artificially inseminated during the latter half of estrus (Table 20-3). Many of the embryo deaths occur early enough for the cows to return to estrus at the normal time. However, a significant number of embryos die between 20 and 40 days. When dealing with a cross section of

TABLE 20-3.   Fertilization, embryonic survival, and embryonic mortality rates of bulls with histories of either low or high fertility in artificial breeding

| Histories of bulls in A.B. | 3-day Slaughter data | | 33-day Slaughter data | | Causes of repeat breedings | | |
| --- | --- | --- | --- | --- | --- | --- | --- |
| | No. obser. | % fert. eggs | No. obser. | % normal embryos | % eggs not fert. | Est. % embryonic mortality | % breeding failure caused by embry. mort. |
| High | 29 | 96.6 | 29 | 86.1 | 3.4 | 10.5 | 75.5 |
| Low | 26 | 76.9 | 26 | 57.7 | 23.1 | 19.6 | 46.3 |
| High–low | | 19.7 | | 28.4 | 19.7 | 9.1 | 29.2 |

From Bearden, PhD Dissertation, Cornell University, 1954.

cattle from several herds one would expect that the difference between predicted and actual pregnancy rate could be caused by embryonic mortality. Some of the positive samples may have been due to nonpregnant cows with long but normal estrous cycles and some may have been due to errors in sampling (wrong cow or wrong time in the cycle). A method that could be used to accurately predict 60-day pregnancies utilizing milk samples taken at 21 to 24 days would be highly desirable. However, due to the problems related to embryonic mortality the results reported here may be as close as we can come.

The greatest shortcoming of the milk progesterone assay method is that it employs radioisotope assay techniques which must be conducted in a central laboratory having expensive isotope counting equipment. At present such laboratories can only be justified in areas of high livestock population. About 5 to 7 days are required for the results to be available to the cattlemen. This delay may further decrease the usefulness of the test, but it still provides earlier detection of some open cows than can be determined with rectal palpation.

Research is under way to develop an enzyme immunoassay for milk progesterone that may be simple enough to be conducted on the farm. The procedure employs enzyme labeled progesterone which competes with the progesterone in the milk sample for binding sites on an antiprogesterone antibody bonded to test paper. If the test can be perfected, it might be practical to use it as a supplement to the regular estrus checking program. Samples could be assayed twice weekly beginning on day 18 post-breeding and continued until the cow either returned to estrus or could be determined pregnant by rectal palpation. The procedure would have to be simple, fast and inexpensive to be practical.

Beef cattle managers possibly would prefer to take blood samples rather than milk samples for progesterone assay. It will be up to the manager to determine whether the early pregnancy detection is worth the time, effort, and expense.

**20-2**
**Ewe**

20-2.1
Progesterone
assay

Research utilizing the progesterone assay (milk or blood) method described in the preceding section has not been reported for the ewe. However, the authors feel that the procedure may have more significance for the ewe than for the cow in that rectal palpation cannot be used on the ewe. Additional research will have to be conducted before an answer will be available.

20-2.2
Ultrasonic
sound

Ultrasonic sound equipment has been available for many years and has been used to make a number of different determinations on live animals. The technique is based on detection of a difference in acoustical impedance between tissues or structures contained in the body. Equipment and techniques utilized in human obstetrics were adapted for use on the ewe. Wool was shorn from the abdomen from flank to flank 8 to 10 cm in front of the udder. The ultrasonic probe was then passed along this line aimed toward the reproductive tract. Positive diagnosis was based on detecting a fetal pulse and swishing of the umbilical cord. Unfortunately, the method is only effective in diagnosing pregnancy in ewes 100 or more days post-breeding.

A newer probe which is small enough to be inserted into the rectum of the ewe will diagnose nonpregnant ewes between 20 and 80 days post-breeding with 88% accuracy. Positive pregnancy diagnoses for the same period were 84% accurate. Each ewe was checked only once. Because of the relatively short breeding season pregnancy information must be available early if it is going to be of value (Figure 20-10).

FIGURE 20-10. A version of the ultrasonic sound equipment used to diagnose pregnancy in the ewe. Note wool has been clipped from abdomen in front of udder.

**20-2.3**
**Vaginal biopsy**

Vaginal biopsy is a method that has shown up to 97% accuracy in diagnosing pregnancy in the ewe. The method is not one that can be used on the farm. Biopsies must be obtained surgically and processed for histological preparation. Vaginal epithelium of the nonpregnant ewe contains ten to twelve layers of cells. The surface layers are squamous while the deeper layers are polygonal. The epithelium of the pregnant ewe contains fewer layers and these are columnar and cuboidal rather than squamous and polygonal. The test can be conducted from 40 days post-breeding to term.

**20-3**
**Mares**

The most practical and earliest determination of pregnancy in mares is done by rectal palpation. The most practical time for palpating the mare is 35 days post-breeding. While the procedures are very similar to the rectal palpation method in cows, some differences will be described.

**20-3.1**
**Rectal palpation**

*Restraint.* Some mares may be palpated without restraint. But being kicked by a mare may be serious if not fatal, so all mares should be properly restrained before being palpated. A properly designed breeding stock that protects the examiner may be used (Figure 20-11). A number of commercial as well as improvised breeding hobbles may be used. Any such equipment should provide protection for the examiner, should not cause the mare to become excited and should be easy and safe to apply and remove. A nose twitch may be effective with some mares but causes additional excitement in others. Very nervous mares may have to be sedated or tranquilized before the application of restraint equipment. The restraint should also prevent the animal from moving sideways to prevent injury to the elbow.

*Lubrication.* The rectal mucosa of the mare is much dryer than the mucosa of the cow. Proper lubrication is essential for successful palpation and to prevent damage to the rectum.

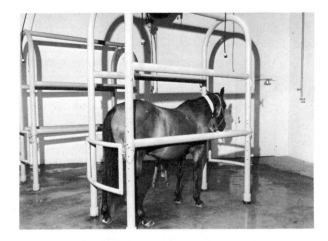

FIGURE 20-11.   Mare being restrained in a stock for palpation.

**20-3.1a Structures to be palpated**   The ovaries, rather than the cervix, serve as a landmark or base of operations in the mare. A systematic examination is begun with location of one of the ovaries. The right ovary is easier to reach and locate when the left hand is used by the examiner. In nonpregnant and early pregnant mares the ovaries are located 5 to 10 cm anterior to the upper third of the pelvic arch. They can be recognized by their distinct oval and irregular form and rather firm consistency. Detailed examination of the ovaries does not contribute information of significance for pregnancy diagnosis.

*Uterus and Its Contents.* The diagnostic decision in the mare is based almost entirely on the size of the uterus and its contents. The mare's uterus is bipartite with the horns joining the body of the uterus almost perpendicularly resulting in a T-shaped structure. The horns are slightly funnel shaped and are 10 to 16 cm in length. The body of the uterus measures 15 to 20 cm in length with a width of 4 to 6 cm. After locating an ovary the broad ligament is followed down to the top of the uterine horn. A systematic examination of both horns and the body of the uterus is necessary in making the diagnosis. The uterus of the open or early pregnant mare will be located in the pelvic region. As pregnancy progresses the uterus and its contents will be pulled into the abdominal cavity.

Fetal membranes and placentomes which are positive signs of pregnancy in the cow cannot be used as criteria of pregnancy in the mare because of the diffuse attachment of the placenta. Bulging of the uterus caused by the amnionic vesicle is readily detectable between 30 and 50 days post-breeding. Beyond this point the fetus itself can be palpated.

*Pulse of Pregnancy.* The demand for an increased blood supply to the uterus causes the same enlargement and increased blood supply in the uterine arteries as observed in the cow. Both the utero-ovarian and middle uterine arteries may be palpated and changes may be detected after 150 days of gestation.

*Position of the Ovaries.* The change in the position of the uterus involves the broad ligaments and the ovaries. As the uterus descends into the abdominal cavity its weight pulls the broad ligament, which in turn moves the ovaries downward and forward. By the sixth to seventh month of gestation the ovaries may be level with the pelvic brim.

**Palpation at 35 Days.** This stage is characterized by the appearance of the amnion which is about the size of a golf ball in one of the uterine horns. The remainder of the uterus does not have the fluid-filled feeling as in the cow at this stage. The uterine muscles have good tone.

**Palpation at 42 to 45 Days.** The amnion at this stage attains a slightly oval form and measures 5 to 7 cm in length and approximately 5 cm in diameter. Its position has reached the junction of the horn and body of the uterus (Figure 20-12).

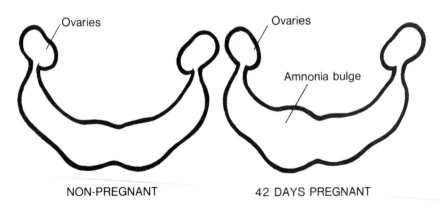

Ovaries                    Ovaries

Amnonia bulge

NON-PREGNANT            42 DAYS PREGNANT

FIGURE 20-12.   Diagram of a mare's uterus showing an amnionic bulge typical of a 42-day pregnancy.

**Palpation at 60 Days.** Approximately half of the amnionic vesicle is located in the body of the uterus. It has assumed the shape of a football and is 12 to 15 cm long and 8 to 10 cm in diameter.

**Palpation at 90 Days.** The entire body of the uterus is involved at this stage. The enlarged area measures approximately 20 to 25 cm in length and about 12 to 16 cm in diameter. About half of the body of the uterus will be pulled over the brim of the pelvis. The fetus is easily palpated at this stage.

**Palpation at 3 to 5 Months.** The entire uterus will be pulled into the abdominal cavity. The ovaries will be pulled down and forward. The fetus can still be palpated.

**Palpation at 5 to 7 Months.** The descent of the uterus continues and will be completed toward the end of this period. The ovaries continue to be pulled downward and forward and toward the end of the period will be approximately even with the floor of the pelvis. The fetus can be palpated in most mares. The uterine arteries may be palpated as described for the cow.

**Palpation at 7 Months to Parturition.** The fetus is easily palpated during this period.

**20-3.1b Differential diagnosis**   Pathological conditions which might be confused with pregnancy are very rare in the mare. Pyometra, although rare, is relatively easy to distinguish from pregnancy. The uterus can be retracted and the absence of a fetus established. Retention of a dead fetus in the form of mummification has not been encountered in the mare.

**20-3.2**
**Hormone assay**     Gonadotrophic hormones (PMSG) appear in the blood about 40 days post-breeding. Peak levels appear between 50 and 120 days and then begin a gradual decline. A bioassay test using immature female rats (25 to 40 days of age) has been developed for detecting the presence of PMSG. These rats are

injected with serum from a mare and killed 72 hours later. The ovaries are examined for follicular growth. An immunologic test kit is also available and can be purchased. Both tests give best results when performed between 50 and 100 days post-breeding.

20-3.3
Progesterone
assay

Recent research has revealed that mare's milk can be assayed for progesterone to diagnose pregnancy. Since the mare is usually lactating at the time of breeding, milk samples can be obtained. Progesterone assay on blood may also be a possibility. Additional research needs to be performed in this area.

**20-4
Sows**

The ultrasonic sound technique will be the only method described for detecting pregnancy in the sow. The equipment and technique have been developed that approach 100% accuracy (Figure 20-13). The tip of the transducer is placed in contact with the lower flank of the standing sow about 5 cm posterior to the navel and lateral to the nipple line. An easily recognized band of echoes is obtained from a depth of 15 to 20 cm in pregnant sows to contrast with echoes from a depth of only 5 cm in nonpregnant sows. Accurate results have been reported from 30 to 90 days of gestation. In field work, determinations at 35 days have been reported accurate and provides information at a time when it can be of most value. If sows are checked for pregnancy at 35 days those not pregnant can be scheduled for rebreeding or slaughter resulting in considerable savings in feed cost.

Progesterone assay on blood may be a possibility and may be more practical than with beef cows or ewes.

FIGURE 20-13.   Sow being checked for pregnancy with ultrasonic equipment.

**Suggested reading**

Bearden, H. J., W. Hansel and R. W. Bratton. 1956. Fertility and embryonic mortality rates for bulls with histories of either low or high fertility in artificial breeding. *J. Dairy Sci.*, 39:312.

Estergreen, V. L. 1978. A simplified test for milk progesterone and pregnancy testing. *The Advanced Animal Breeder*, 24:10. NAAB, Columbia, Mo.

Foote, R. H. 1979. Hormones in milk that may reflect reproductive changes. *Beltsville Symposia in Agricultural Research. 3. Animal Reproduction*, p. 111. Allanheld, Osmun and Co. Publisher; Allsted Press, a division of John Wiley and Sons.

Hafez, E. S. E. 1974. *Reproduction in Farm Animals*. (3rd ed.) Lea and Febiger.

Richardson, C. 1972. Pregnancy diagnosis in the ewe: a review. *Vet. Rec.* 90:264.

Sorensen, A. M., Jr. and J. R. Beverly. 1968. Determining pregnancy in cattle. *Texas Agr. Ext. Ser. Bull.* 1077.

Zemjanis, R. 1970. *Diagnostic and Therapeutic Techniques in Animal Reproduction*. (2nd ed.) The Williams and Williams Co. Baltimore, Md.

Zemjanis, R. 1974. Pregnancy diagnosis. *Reproduction in Farm Animals*. (3rd ed.) ed. E. S. E. Hafez. Lea and Febiger.

# Chapter

# 21 Environmental management

Most animal managers consider adverse stress of any nature to be undesirable in relation to reproductive efficiency. *Stress* can be defined as any environmental change (i.e., alteration in climate or management) that is severe enough to elicit a behavioral or physiological response from the animal. With the exception of high-temperature stress, the individual response of animals to adverse stress varies greatly. Such stresses as cold, transportation, crowding, and simple changes in management routine have sometimes but not always reduced reproductive efficiency. Typical responses to these stresses included altered length of the estrous cycle, delayed ovulation and in some cases, suppressed ovulation. All of these responses can reduce reproductive efficiency.

A number of adverse effects can be associated with heat stress. Those discussed in earlier chapters included delayed puberty in both sexes and production of low quality semen or even temporary sterility in males. In females, irregular estrous cycles, short periods of estrus, more silent estrus, and delayed or suppressed ovulation frequently occur. If the ambient temperature is high enough to elevate rectal temperatures of females by 1 to 2°C, marked reductions in conception rate are seen (Figure 21-1). The lowered conception may result from either fertilization failure or early embryonic mortality. In sheep and beef cattle heat stress during gestation has resulted in smaller and sometimes dwarfed offspring.

Summer conditions in many parts of the world are severe enough to adversely affect reproduction. Ambient temperatures above 30°C will usually lower conception rate (Figure 21-2). Higher daytime temperatures can be tolerated if the nights are cool (<18°C). High ambient temperature is more detrimental if the relative humidity is also high. In a study conducted in Florida the five climatic factors with the strongest influence on conception in dairy cattle were (1) maximum temperature day after insemination, (2) rainfall day of insemination, (3) minimum temperature day of insemination,

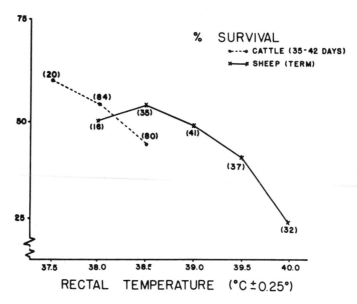

FIGURE 21-1.   Influence of rectal temperatures at or near the time of mating on pregnancy rate in sheep and cattle. Rectal temperatures were taken for sheep at the time of mating and 12 hours after insemination for cattle. The numbers in parenthesis represent the number of observations. (Ulberg and Burfening. 1967. *J. Anim. Sci.*, 26:571.)

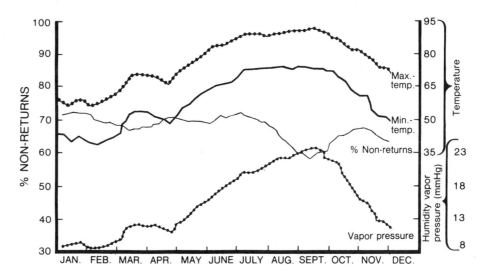

FIGURE 21-2.   Seasonal effect on fertility in dairy cattle in Louisiana using natural mating. (Johnston and Branton. 1953. *J. Dairy Sci.*, 36:934.)

(4) solar radiation day of insemination, and (5) minimum temperature day after insemination. Other researchers have identified the time from insemination to a few days thereafter as the most critical period to conception. Fewer pregnancy losses through embryonic mortality have occurred when females have been subjected to heat stress after the cleaving embryo has reached the uterus.

**21-1 Physiological relationship of environmental stress to reproduction**

Most researchers believe that general stress exerts its influence through the endocrine system (Figure 21-3). The mechanism is still being debated, but an involvement of the hormones of the adrenal cortex has received considerable attention. It is known that stress will cause the release of ACTH from the anterior pituitary which, in turn, stimulates release of cortisol and other glucocorticoids. Glucocorticoids inhibit the release of LH. Therefore, if an animal is under stress during a critical period of the estrous cycle (late proestrus or estrus), a glucocorticoid induced suppression of LH is likely to either delay or prevent ovulation and may reduce libido in males.

Heat stress will cause the release of ACTH and glucocorticoids, also (Figure 21-1). As the animal adapts to high temperature, glucocorticoids will return to near prestress levels, but this usually requires several days. Therefore, with the onset of heat stress the reproductive response will be similar to that seen in other types of stress. In addition to its effect on the adrenal

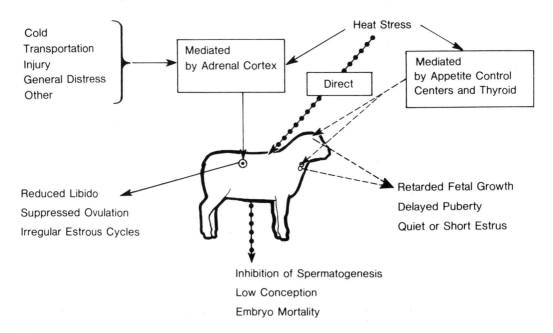

FIGURE 21-3. Reproductive phenomena impaired by adverse stress and possible physiological mechanisms of action.

gland, heat stress depresses appetite and reduces thyroid activity with a resulting lower basal metabolic rate. There is little adaptation to this as depressed appetites and lower basal metabolic rates are seen throughout the summer. Heat stress differs in this regard because other stresses usually either have little effect or else stimulate appetite and thyroid activity. Reduced feed intake and lower thyroid output contribute to temperature induced delays in puberty and probably to shorter, quieter estrus and birth of smaller offspring.

There is also evidence that heat stress has a direct effect on reproductive efficiency in addition to its effect through the endocrine system. The interference with spermatogenesis appears to be a direct effect of an elevated temperature on the germinal epithelium. If a female's internal temperature is elevated, there is evidence that damage can occur to (1) the spermatozoa while capacitating in the female tract, (2) the oocyte after ovulation, and/or (3) the embryo during the early cleavages. The damage to the gametes does not interfere with fertilization, but results in embryonic mortality before placentation. These effects, combined with the endocrine changes which occur during heat stress, account for the more pronounced effect of heat stress on reproduction than is seen with other stresses.

In understanding the variable reproductive response to stress, one must understand that animals will adapt to specific stressors. Animals maintained in a cold environment will usually reproduce normally. On the other hand, adapting to cold stress may make them more susceptible to other stressors. If, in addition to cold stress, estrous females are subjected to wind and rain or are transported to a new location for insemination, the physiological mechanisms that interfere with reproduction may be triggered. It is important to recognize that stress may interfere with reproduction. Frequently, simple modifications in management will lessen the likelihood of stress occurring around the time of estrus and insemination.

**21-2 Thermo-regulation**

Management to alleviate undesirable stress involves recognizing and eliminating the cause of the stress. Whatever the stress, the most critical period extends from late proestrus through metestrus. Since reduced reproductive efficiency is more predictable with heat stress and may include problems in detection of estrus, conception, and fetal growth, a more basic understanding of the animal's response to heat stress is needed. This will help the animal manager adopt practices to increase reproductive efficiency during the summer.

*Thermoregulation* is defined as the means by which an animal maintains its body temperature (Figure 21-4). It involves a balance between heat gain from metabolism and the environment and heat loss through metabolism and to the environment. Basically, the heat gained from metabolism is that necessary for living, production, and reproduction. Man has limited control over either heat gain or heat loss from metabolism. This is regulated within the animal, principally through the hypothalamus and related systems. Therefore, the heat gained from the environment must be offset by heat loss to the environment. If heat gained from the environment exceeds heat loss, body temperature will rise. In addition, physiological adjustments will occur to reduce metabolic heat gain. These physiological adjustments result in reduced productivity. Also, the stress response (section 21-1) and increased body temperature will lessen the likelihood of gestation with its associated increment of metabolic heat. To maximize reproductive efficiency during the summer, measures must be taken which reduce heat gain and/or facilitate heat loss to the environment.

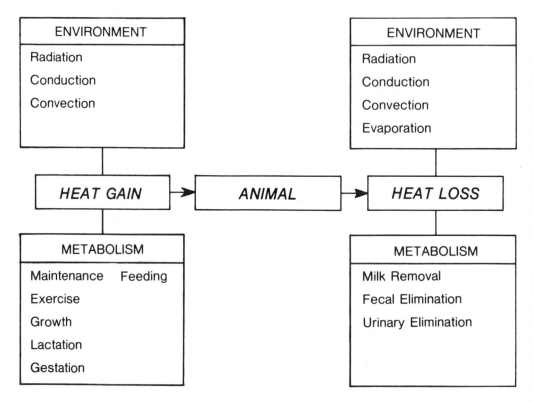

FIGURE 21-4. Increments of heat gain and heat loss which are balanced to contribute to thermoregulation in an animal. (Fuquay. *J. Anim. Sci.* (In Press).)

**21-3
Modification of
summer
environments
to reduce
stress**

During the day most of the heat gained from the environment comes directly or indirectly from solar radiation. Solar radiation is absorbed when animals are exposed to direct sunlight. Indirect radiation comes from objects which absorb and re-emit radiation (e.g. metal buildings or paved lots) and reflected radiation from light colored surfaces and clouds. Radiation that strikes objects on the earth is either absorbed or reflected. Much of that which is reflected will be lost to a clear sky. If the skies are partly cloudy, clouds will deflect some of the reflected radiation back to the earth adding to the radiation problem.

*Shades* provide protection from solar radiation and will reduce daytime heat stress during the summer. Both natural shading provided by trees and artificial shades constructed from a variety of materials are beneficial. The efficiency of metal roofs as *radiation shields* can be improved by painting them white, insulating the underside, or using a sprinkler system to keep them wet. (See Table 21-1).

Most of the heat loss during the summer is from evaporation of moisture from skin or lungs. Even though cows and sheep are not considered sweating animals, evaporation of natural secretions from their skin surfaces is a major avenue of heat loss during the summer. Pigs are less efficient sweaters, but compensate by lying in water to keep skin surfaces wet if given the opportunity.

Cooling by evaporation is facilitated by reducing the dew point temperature around the animals. This can be accomplished by lowering the relative humidity or the air temperature. Reduction of relative humidity is quite costly, but use of the evaporative cooler to lower the air temperature has proven to be economically feasible for dairy farms in hot arid regions and has improved reproductive efficiency. Increasing the movement of air around animals will greatly enhance cooling by evaporation if the dew point temperature of the air is lower than that of the skin. Increased air movement can facilitate heat loss by evaporation during both the day and night. Further, increased air movement will be of benefit in humid regions where mechani-

TABLE 21-1.  Options available to animal managers that reduce heat stress and are likely to be economically feasible during the summer

| | |
|---|---|
| *Radiation shields** | Natural shade |
| | Artificial shade |
| *Mechanical cooling* | Fans and/or sprinklers |
| | Evaporative coolers (arid regions) |
| *Dietary changes* | Low fiber diets (ruminants) |
| | Cooled water (ruminants) |

*Radiation shields are only protective during the day. At night radiation shields inhibit radiational cooling.

cal evaporative coolers are inefficient. Direct sprinkling of cattle and pigs has been beneficial in arid regions, but is less so in humid regions.

Radiational cooling is an important means of heat loss at night. Animals will radiate to the cooler night sky and lose body heat. The same radiation shields that reduce radiant heat gain during the day will inhibit radiant heat loss at night. During summer nights, animals need to have access to comfortable areas that are not covered by radiation shields.

The economic benefit from environmental modifications in terms of reproduction, meat, and/or milk production must be balanced with the cost of providing the added comfort. Shades with natural cross-ventilation provide sufficient comfort and can be justified in most hot environments. If additional cooling is needed, the economic feasibility becomes marginal. Some farm managers have limited the use of artificially cooled environments to the time between parturition and first insemination with good results. Cooling for this period has reduced postpartum interval to resumption of the estrous cycle and improved conception rate. Cooling for a more limited period, around the time of estrus and insemination, may be beneficial if care is taken to insure that movement to the cooled quarters does not cause other physical or psychological stress that is detrimental to conception. Limited cooling is only beneficial when applied to the female. If the internal temperature of the testes is elevated sufficiently to interfere with spermatogenesis, several weeks will be needed before recovery of full fertility.

## 21-4
## Other management considerations in hot environments

One means of avoiding reduced conception rates during the summer is to breed animals during other seasons. This is done to a large degree in beef herds, with spring as the heavy breeding season, and dairy herds, when most cows are bred in late fall and early winter. This avoids the most troublesome reproductive period as related to heat stress, but does not deal with the problem of heat stress on fetal growth. The breeding season of sheep starts during the summer and marketing systems for swine favor farrowing at a uniform rate throughout the year. For sheep and pigs, the alternative of avoiding the summer breeding season is not desirable.

Part of the reduced conception associated with summer breeding can be avoided through artificial insemination by using frozen semen collected from males during other seasons. This alternative is available only in those species where use of frozen semen has reached commercial practicability. It is not yet available in some species where there is great need.

Providing cooled water for drinking has benefited ruminants. Care must be taken because water intake will be reduced if the water is more than 10°C below ambient temperature. Providing low fiber, high energy diets during the summer will also benefit ruminants. The heat of digestion of such diets is lower than for high fiber diets, thus reducing metabolic heat production.

In cows it has been observed that periods of estrus average 10 to 12 hours in hot environments compared to an average of 18 hours for cows in cool environments. Managers in hot environments must check for estrus more frequently with one check coming at dawn. More cows commence with estrus during the night than during the day. Also, they are more likely to show physical signs of estrus during the coolest part of the day. If the first morning check is delayed by 2 to 3 hours, some cycling cows will be missed. More frequent detection is needed to assure that fewer cows are missed. It is also important so that the time of insemination can be synchronized with late estrus to give a higher probability of conception.

**Suggested reading**

Alliston, C. W., B. Howarth and L. C. Ulberg. 1965. Embryonic mortality following culture *in vitro* of one and two-cell rabbit eggs at elevated temperatures. *J. Reprod. Fert.*, 9:337.

Brown, D. E., P. C. Harrison, F. C. Hinds, J. A. Lewis and M. H. Wallace. 1977. Heat stress effects of fetal development during late gestation in the ewe. *J. Anim. Sci.*, 44:442.

Burfening, P. J. and L. C. Ulberg. 1968. Embryonic survival subsequent to culture of rabbit spermatozoa at 38 and 40°C. *J. Reprod. Fert.*, 15:87.

Dutt, R. H., E. F. Ellington and W. W. Carlton. 1959. Fertilization rate and early embryo survival in sheared and unsheared ewes following exposure to elevated air temperatures. *J. Anim. Sci.*, 18:1308.

Fallon, G. R. 1962. Body temperature and fertilization in the cow. *J. Reprod. Fert.*, 3:116.

Gangwar, P. C., C. Branton and D. L. Evans. 1965. Reproductive and physiological response of Holstein heifers to controlled and natural climatic conditions. *J. Dairy Sci.*, 48:222.

Gwasdauskas, F. C., C. J. Wilcox and W. W. Thatcher. 1975. Environmental and management factors affecting conception rate in a subtropical climate. *J. Dairy Sci.*, 58:88.

Hall, J. G., C. Branton and E. J. Stone. 1959. Estrus, estrous cycles, ovulation time, time of service and fertility of dairy cows in Louisiana. *J. Dairy Sci.*, 42:1086.

Stott, G. H., F. Wiersma and J. M. Woods. 1972. Reproductive health program for cattle subjected to high environmental temperatures. *J. Amer. Vet. Med. Assoc.*, 161:1339.

Thatcher, W. W. 1974. Effect of season, climate and temperature on reproduction and lactation. *J. Dairy Sci.*, 57:360.

Ulberg, L. C. 1958. The influence of high temperature on reproduction. *J. Heredity*, 49:62.

Ulberg, L. C. and P. J. Burfening. 1967. Embryo death resulting from adverse environment on spermatozoa or ova. *J. Anim. Sci.*, 26:571.

Vincent, C. K. 1972. Effect of season and high temperature on fertility in cattle. *J. Amer. Vet. Med. Assoc.*, 161:1333.

# Chapter

# 22 Nutritional management

Lack of proper nutrition can reduce reproductive efficiency (Table 22-1). Gross deficiencies or excesses can be easily recognized and steps taken to correct the problem. Other imbalance in nutrition can be subtle and difficult to recognize. Diagnosis is sometimes difficult because reproductive symptoms associated with many deficiencies in nutrition are similar to symptoms caused by other disorders. It should also be recognized that nutrition is often a scapegoat, being falsely blamed for problems in reproduction that are caused by infectious organisms or deficiencies in management.

There are no magic nutritional formulae which insure efficient reproduction. If diets for animals are sufficient to meet their requirements as established by the National Research Council, nutrition is not likely to limit reproduction. Most nutrition related reproductive problems result from either neglect or an over estimation of the nutritional value of the feedstuffs used in formulating diets. Under range conditions, reduced reproductive efficiency has been reported for beef cows during drought years when forage has been sparse.

The objective of this Chapter is to discuss those dietary components that may reduce reproductive efficiency and to give general guidelines on the nutritional management of animals during different phases of reproduction.

## 22-1
## Nutritive components

Nutritional factors needed for successful reproduction are the same as those needed for maintenance, growth, and lactation (Figure 22-1). They include energy, protein, vitamins, and minerals. A deficiency or excess of any of these components which is serious enough to affect reproduction will also affect other physiological functions.

## 22-1.1
## Energy

Many nutrition-related problems in reproduction are due to either underfeeding or overfeeding energy. Energy is unique because in terms of length of reproductive life overfeeding is more detrimental than underfeed-

TABLE 22-1. Nutrient related abnormalities in reproduction*

| Nutrient | Reproductive Disorder |
| --- | --- |
| Energy excess | Low conception, abortion, dystocia, retained placentae, reduced libido |
| Energy deficiency | Delayed puberty, suppressed estrus and ovulation, suppressed libido and spermatozoa production |
| Protein deficiency | Suppressed estrus, low conception, fetal resorption, premature parturition, weak offspring |
| Vitamin A deficiency | Impaired spermatogenesis, anestrus, low conception, abortion, weak or dead offspring, retained placentae |
| Vitamin D deficiency | Defective skeletal development, rickets |
| Calcium deficiency | Skeletal defects, reduced viable young |
| Phosphorus deficiency | Anestrus, irregular estrus |
| Iodine deficiency | Impaired fetal growth, irregular estrus, retained placentae |
| Selenium deficiency | Retained placentae |

*Compiled from literature.

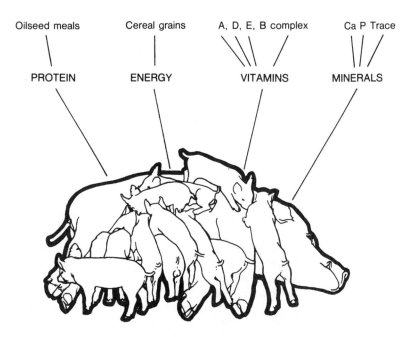

FIGURE 22-1. Balanced nutrition is needed for good reproductive efficiency.

ing. Overfeeding results in excess fatness with fat infiltrating the liver and sometimes the reproductive organs. Such animals have less resistance to infectious diseases and other stress. Also, lower conception rate, abortion, dystocia, and more frequent retention of the placenta have been reported. Over-conditioned males are more likely to have foot and leg problems which reduce libido (see Chapter 11). Extreme fatness is more likely to occur in dairy cows which are fed too liberally during late lactation and in meat animals which have been fattened for livestock shows.

While overfeeding may be more serious, underfeeding occurs more frequently. Underfeeding of energy will delay puberty in both sexes, reduce production of spermatozoa in growing males after puberty, reduce ovulations per estrus in growing, polytoccus females, cause weak or silent estrus, cause or extend periods of anestrus, and reduce libido in males. Some have reported lower conception rates in underfed cattle, but the primary cause of lower calf crops for underfed beef cattle appears to be an extended postpartum anestrus rather than low fertilization rate.

The data in Table 22-2 show the effects of low, medium, and high levels of nutrition on the age and weight at first calving, services per conception, birth weight of calves, and calving difficulty for Holstein heifers. The medium level of nutrition was recommended and even though these heifers were 3.5 months younger than the low group at calving, they weighed 99 kg more and experienced approximately half the calving difficulty. Birth

TABLE 22-2. Influence of plane of nutrition on reproductive performance of Holsteins

| Measure | TDN intake (percent of normal) | | |
| --- | --- | --- | --- |
|  | 62 | 100 | 146 |
| Number of heifers | 33 | 34 | 34 |
| Age at first estrus, months | 20.2 | 11.2 | 9.2 |
| Weight at first estrus, kg | 303 | 265 | 277 |
| Conceived to first mating, % | 79 | 68 | 58 |
| Matings per first conception, No. | 1.55 | 1.41 | 1.48 |
| Age at first calving, months | 32.0 | 28.5 | 27.9 |
| Weight at first calving, kg | 384 | 483 | 548 |
| Weight of first calf, kg | 36 | 39 | 41 |
| Requiring help at calving, % | 45 | 26 | 24 |
| Delivering living calves, % | 87 | 88 | 94 |
| Matings per second conception, No. | 1.71 | 1.76 | 2.09 |
| Matings per third conception, No. | 1.90 | 1.64 | 1.90 |
| Culled for sterility, % | 6 | 12 | 20 |

From Maynard and Loosli. *Animal Nutrition.* McGraw-Hill Book Co. 1969.

weights of calves were not greatly affected by weight of dams. Services per conception were similar for all groups. The medium level group could have been bred 3 months earlier without additional energy intake.

Major sources of energy include cereal grains, grain by-products, and high quality forages. For ruminants legume hay, corn silage, summer and winter annual grasses (grazed or fed fresh), and perennial grass-legume pastures during the spring are good sources of energy. Perennial grasses are frequently overrated during the summer. This is particularly true in subtropic and tropic regions. These grasses remain green, but growth rate and soluble carbohydrates are reduced while the fibrous, structural portion increases. They cannot be digested quickly enough to meet the energy requirements of growing or lactating ruminants. This deficiency may not be readily apparent to a casual observer, but can result in delayed puberty or extended postpartum anestrus. Sorghum silage is also frequently overrated as an energy source. Most of the carbohydrate in green sorghum plants is in the form of soluble sugars. When grazed or fed freshly cut, it is a good source of energy. When ensiled, the soluble sugars are lost during fermentation in the silo. By contrast, most of the carbohydrate in corn silage is starch, which is not lost during fermentation.

## 22-1.2 Protein

Diets deficient in protein have resulted in weak or silent periods of estrus, cessation of estrus, repeat breeding, fetal resorption, and birth of premature and/or weak offspring. In practice, diets low in protein are also frequently low in energy. For pigs, the quality of the protein must be considered, since adverse effects have been observed when one or more amino acids are limited. Diets deficient in a single amino acid have slowed growth rate resulting in delayed puberty, extended postpartum anestrus, and abnormal fetal growth patterns.

Best sources of protein are high protein concentrates such as soybean, cottonseed, and peanut meals and legumes such as alfalfa or clover. Both annual and perennial grasses can provide a good source of dietary protein with adequate moisture and liberal fertilization with nitrogen at earlier stages of maturity. Haylage and low moisture silage are frequently overrated because the protein digestibility is decreased by excess heating in the silo. A similar problem exists for heat damaged hay.

Use of urea and other sources of nonprotein nitrogen in diets of ruminants will not adversely affect reproduction if fed as recommended.

## 22-1.3 Vitamins

It is likely that all vitamins needed for growth and maintenance are also needed for reproduction. However, because of the availability in the usual animal feedstuffs, few have become sufficiently deficient to adversely affect reproduction.

A deficiency in vitamin A can adversely affect reproduction in both males and females. It is necessary for the integrity of the germ cells in the seminiferous tubules in males. A deficiency can reduce or even stop spermatogenesis. In females, a deficiency of vitamin A has resulted in an array of problems including anestrus, repeat breeding, abortions, weak or dead offspring, and retained placentae. Vitamin A can be stored in the liver. Therefore, a body reserve is available to prevent problems during short periods when the diet is deficient.

Fresh green forages, air cured alfalfa hay, green silage, and purchased concentrate supplements are the best dietary sources of vitamin A. Vitamin A is destroyed by high temperature or by sunlight. Therefore, haylage, low moisture silage, heat damaged hay, and sun cured hay are not good sources. Most vitamin A deficiencies in ruminants occur when the forage is from one of those sources. When the vitamin A content of the forage is in doubt, addition of a vitamin A supplement to the concentrate is simple and secure insurance against a vitamin A deficiency.

As a result of early research in rats, vitamin E was called the "antisterility vitamin." In female rats a deficiency resulted in death and resorption of feti. In male rats degeneration of the testes sometimes resulted in permanent sterility. Efforts to demonstrate a need for supplemental vitamin E for reproduction in other species have not been successful. The availability of vitamin E in cereal grains and forages make a deficiency of this vitamin unlikely. Salesmen have promoted the use of vitamin E and wheat germ oil, which is high in vitamin E, as a cure for both real and imagined reproductive problems. However, the need for this vitamin as a supplement has not been demonstrated.

Vitamin D is of interest in reproduction because of its role in absorption and retention of calcium and phosphorus. Principal effects of vitamin D deficiency have been on development of the skeletal system of the fetus. Rickets are common when a deficiency has existed during gestation. Vitamin D deficiencies do not occur unless animals are deprived of sunlight. Sun cured hay and supplements which can be added to concentrate feeds will also provide this vitamin. Care must be exercised when supplementing diets with vitamin D, especially when animals are exposed to adequate sunlight. Vitamin D toxicity has occurred in both humans and farm animals from an excess of the vitamin.

Most other vitamins are either synthesized by the animals or are available in adequate quantities in the usual feedstuff to prevent deficiencies. Vitamin $B_{12}$ is not available from plant sources. It is synthesized in the rumen of ruminants and provided through the addition of $B_{12}$ supplement or meat products to the diet of simple stomached animals.

**22-1.4**
**Minerals**

Requirements of most minerals are increased, per unit of body weight, by gestation, lactation, and growth. While gross deficiencies may reduce efficiency of reproduction, such deficiencies seldom occur in field conditions. Mineral mixes containing calcium, phosphorus, iodized salt, and trace minerals are recommended in the nutritional management of animals for meat and milk production. If these minerals are supplied in sufficient quantities and ratios to meet the requirements for meat and milk production, it is doubtful that reproduction will be affected. If mineral mixes are not provided, the mineral content of natural feedstuffs must supply mineral needs. Reproductive problems have resulted in areas where deficiencies of specific minerals in the soil have reduced that found in natural feedstuffs.

Reduced calf crops have been reported in phosphorus-deficient areas of the world. In a phosphorus-deficient region of South Africa, supplementation with bone meal to provide phosphorus resulted in a 29% improvement of calf crop (80% vs 51%). Low calf crop in phosphorus-deficient cows appears due to irregular estrous patterns and periods of anestrus. Diets deficient in phosphorus will depress appetite which could contribute to anestrus and delayed puberty.

Calcium deficiency is seldom a problem in ruminants. Legumes are high in calcium. In addition, animals have physiological control over both absorption and excretion of calcium. Calcium-deficient swine rations have reduced the number of viable young per litter and in some cases caused intrauterine death. The effect of calcium deficiency on the development of the fetal skeletal system might be more severe, except that the fetus can draw minerals from the skeletal system of the mother for development of its own system. If the calcium reserves of the mother are not replenished between gestations, skeletal defects may occur in later gestations. This is a potential problem but seldom occurs. Most defects in the development of the skeletal system of the fetus are due to a lack of vitamin D, which has been discussed.

Iodine deficiency has been reported to cause a number of disorders in reproduction. They include delayed development of the reproductive tract, irregular estrus, impaired fetal growth, and retained placentae. Certain regions are deficient in iodine. In such regions, allowing animals to have access to iodized salt provides sufficient protection.

A higher than normal incidence of retained placentae has been reported in cows in selenium-deficient regions. An injection of selenium or selenium and vitamin E about two weeks before expected parturition will reduce the incidence of retained placentae. High levels of selenium can be toxic but specific adverse effects on reproduction have not been reported.

**22-2 Growing animals**

The primary nutritional concern in females between birth and maturity is to maintain a level of intake and thus growth that will (1) not delay puberty; (2) insure that adequate size be attained at the desired breeding age; and (3) maintain growth during gestation so that parturition is not made more difficult by the impaired size of the mother (Chapters 5 and 19). In growing males, the nutritional regime should be sufficient to (1) not delay puberty and (2) not reduce libido and spermatozoa production after puberty (Chapter

11). The nutritional program is controlled by the animal manager and if not managed properly will extend rather than shorten generation interval. Through nutritional management, the animal manager probably has his greatest control over generation interval.

During periods of growth animals should be maintained on a moderate to heavy plane of nutrition. The goal is to maintain an adequate and steady rate of growth until mature size is reached. The dairy heifer can be used as a model for discussing nutritional needs during growth. The dairy heifer will reach puberty at 8 to 13 months of age at a weight of 160 to 270 kg. Both age and weight vary among breeds. If she is to calve at 24 months of age (the optimum age), she will need to reach her minimum recommended breeding weight (Table 19-1) by 15 months of age. To reach this goal the dairy heifer should gain approximately 0.7 kg per day from birth to 15 months of age. A similar pattern of growth should continue through gestation. A heifer on native forage will not be able to maintain the desired growth rate without supplemental feeding during most of the year. As mentioned previously, young, succulent plants are high in soluble carbohydrates and proteins and will meet these requirements for growing aninals. As these plants become more mature there will be proportionately less soluble carbohydrate and protein with more structural carbohydrates in these plants. Ruminants can digest structural carbohydrates, but the rate of digestion is not high enough to meet the animal's energy requirements for growth plus maintenance. During late gestation the additional requirement to account for the growing fetus must be added to requirements for growth and maintenance.

To assure that the nutritive requirements of the growing animal are met, one must know the requirements for energy, protein, vitamins, and minerals. These have been published by the National Research Council (NRC) for dairy cattle, beef cattle, pigs, sheep, and horses. In addition one must know the nutritive value of the components of the animal's diet. Forages vary greatly in nutritive value depending on amount and kind of fertilization, stage of maturity, season, and methods used in storing and preserving. Therefore, forages used in the diets of ruminants should be subjected to laboratory analysis so that intelligent supplementation can be implemented to meet the nutritive requirements of growing animals. When comparing a maintenance diet with one that is designed for growth, several differences will be noted. Principally, the diet of growing animals will need to be richer in energy, protein, calcium, and phosphorus, since growth involves development of muscles and the skeletal system. Requirements of vitamins and other minerals will be higher also.

## 22-3 Maintaining reproductive efficiency

The importance of nutrition in maintaining reproductive efficiency varies with species. Nutrition seldom limits reproduction in mature, lactating dairy cows. In other farm species, where the management systems frequently result in animals receiving marginal or submaintenance diets during part of the year, limited reproduction related to dietary deficiencies is not uncommon. It is important to remember that nutritive requirements increase during gestation. This is especially true during the last quarter of gestation when supplementation is needed to prevent fetal depletion of maternal nutrient reserves. Depletion of nutrient reserves may not interfere with the ongoing gestation unless the deficiencies are severe. Depletion of maternal reserves is more likely to reduce the number of offspring born to future gestations or delay postpartum estrus and the next gestation.

## 22-3.1
## Flushing

Low nutrition will sometimes cause anestrus in cows and reduce ovulation rate in ewes and sows. Conversely, animals in good condition and on a high plane of nutrition will have more offspring on a herd or flock basis than animals in poor condition. It has also been demonstrated that if animals are in a poor nutritional state and thin condition, supplementing their diet so that they are gaining in weight at time of breeding will increase piglets per litter, multiple births in sheep, and percentage of calf crop in beef cows. Supplementing the diet to have animals in gaining state at time of breeding is called *flushing*. The flushing effect in pigs may be offset by higher embryo mortality if heavy supplementation is continued after breeding. In beef cows the principal effect from flushing appears to be that of bringing animals out of anestrus, so that more have the opportunity to conceive during the next breeding season.

## 22-3.2
## Dairy cows

It has been demonstrated in other species that having animals in a gaining state at time of breeding will improve reproductive efficiency. If dairy cows are to maintain a 12-month calving interval, they must be bred at a time when most are losing weight. They are losing weight because the energy and protein output through milk in early lactation exceeds the energy and protein that can be consumed and utilized from their diet. Management of the dairy cow during the last two months of the gestation is critical. She will not be lactating during that two month period. It is recommended that dairy cows be "turned dry" (milking stopped) 60 days prepartum. Her nutrient needs are those needed for maintenance and for her fetus. The manager should feed her to meet her nutrient requirements and to store reserves for early lactation, while guarding against overconditioning.

The manager should also be aware that changing the diet from that needed during the 60-day nonlactating period before parturition to that desired during early lactation places great stress on the digestive system. During the nonlactating period the diet is typically greater than 95% forage with the remainder concentrate. During early lactation this changes to 60 to 70% concentrate. This change should not be made suddenly. It should be started during the last week of the gestation with daily concentrate gradually increased so that the rumen can adjust to higher concentrate before parturition. Having her on a high level of concentrate as lactation starts will reduce the depletion of nutrient reserves from her body and hopefully reduce digestive disturbances that could lead to ketosis or other disorders. The cow that comes through parturition without sickness will be more likely to maintain the desired frequency in reproduction.

## 22-3.3 Males

Nutritional recommendations for growing males are similar to those for growing females. Moderate to heavy feeding is recommended. Underfeeding will delay puberty. After puberty, underfeeding will reduce libido and spermatozoa production in growing males. The principal concern is to provide enough energy and protein, but vitamins and minerals cannot be overlooked.

The mature male should be restricted to a maintenance diet. Underfeeding for an extended period will reduce libido, which effectually reduces spermatozoa production. However, feeding above the recommended diet for maintenance will not further increase libido or spermatozoa production. It may shorten the reproductive life of the male. (See Chapter 11.)

## Suggested reading

Clark, R. T. 1934. Studies on the physiology of reproduction in the sheep. I. The ovulation rate of the ewe as affected by plane of nutrition. *Anat. Record.*, 60:125.

El-Sheikh, A. S., C. V. Hulet, A. L. Pope and L. E. Casida. 1955. The effect of level of feeding on the reproductive capacity of the ewe. *J. Anim. Sci.*, 14:919.

Foote, W. C., A. L. Pope, A. B. Chapman and L. E. Casida. 1959. Reproduction in the yearling ewe as affected by breed and sequence of feeding levels. I. Effect on ovulation rate and embryo survival. *J. Anim. Sci.*, 18:453.

Hafez, E. S. E. 1959. Reproductive capacity of farm animals in relation to climate and nutrition. *J. Amer. Vet. Med. Assoc.*, 135:606.

Hafez, E. S. E. 1960. Nutrition in relation to reproduction in sows. *J. Agric. Sci.*, 54:170.

Joubert, D. M. 1954. The influence of high and low nutritional planes on the estrous cycle and conception rate of heifers. *J. Agric. Sci.*, 45:164.

Julien, W. E., H. R. Conrad, J. E. Jones and A. L. Moxon. 1976. Selenium and vitamin E and incidence of retained placenta in parturient dairy cows. *J. Dairy Sci.*, 59:1954.

Lashley, J. F. and R. Bogart. 1943. Some factors influencing reproductive efficiency of range cattle under artificial and natural breeding conditions. *Mo. Agric. Expt. Sta. Bull.* 376.

Reid, J. T. 1960. Effect of energy intake upon reproduction in farm animals. *J. Dairy Sci.*, (supp.) 43:103.

Ronning, M., E. R. Beronsek, A. H. Kulman and W. D. Gallop. 1953. The carotene requirements for reproduction in Guernsey cows. *J. Dairy Sci.*, 36:52.

Self, H. L., R. H. Grummer and L. E. Casida. 1955. The effect of various sequences of full and limited feeding on the reproductive phenomena of Chester White and Poland China gilts. *J. Anim. Sci.*, 14:573.

Wiltbank, J. N., W. W. Rowden, J. E. Ingalls and D. R. Zimmerman. 1962. Effect of energy level on reproductive phenomena of mature Hereford cows. *J. Anim. Sci.*, 21:219.

Part

# 5 Causes of reproductive failure

# 23 Anatomical and inherited causes of reproductive failure

A freemartin condition is considered to be a problem in cattle. The condition occurs in the female member of heterosexual twins in which the chorionic membranes of the twins fuse early during embryonic development. This fusion allows an exchange of blood between the twins. In order for the freemartin condition to develop, fusion of the embryonic membranes must occur during the embryonic period prior to the development of the reproductive organs. Between 5 and 10% of the female members of heterosexual twins are not freemartins, presumably because the chorionic membranes fail to fuse or fuse after the reproductive organs differentiate. Freemartin-like conditions have been reported in sheep, goats, and swine. Even though multiple births are much more common in these species, the phenomenon is not considered to be a serious problem.

The freemartin is sterile. The gonads vary widely in appearance and in histological makeup. Some have almost normal ovaries, while others resemble small testes complete with epididymides. The oviducts, uterus, cervix, and most of the vagina fail to develop as tubular structures (Figure 23-1). They often appear as bands of tissue with various abnormal deviations in shape. Varying degrees of development of both the Mullerian and Wolffian ducts contribute to these deviations. The presence of some degree of development of vesicular glands has been frequently reported. The external genitalia generally appear normal. Occasionally the clitoris is enlarged and the tuft of hairs at the tip of the vulva is more prominent. The mammary gland remains rudimentary and frequently can be distinguished from a normal mammary gland by one to two months of age.

The cause of the freemartin condition is still not fully understood. As early as 1916 a hormonal theory was presented. The male gonads develop earlier than female gonads (Chapter 8). This is related to the fact that the testes originate from the primary sex cords while in the female the primary sex cords give way to the development of the secondary sex cords, which in turn develop into ovaries. According to the hormone theory, the developing male testes produce hormones, presumably androgens, which interfere with the development of the female reproductive organs. More recent knowledge casts serious doubt on the hormonal theory. Androgens do not appear to have the ability to interfere with the development of the reproductive system. They also do not appear to have the ability to cause the regression of the Mullerian ducts which frequently occurs in the freemartin. Also the external genitalia which are most sensitive to masculinization generally remain normal.

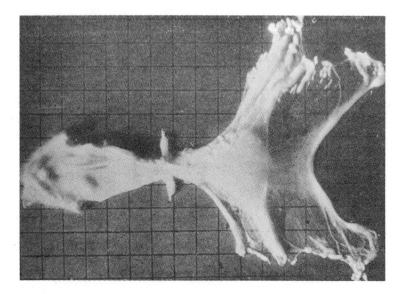

FIGURE 23-1.   Reproductive organs from a freemartin. (From Asdell and Bearden. 1959. *Cornell Ext. Bull.* 737.)

Current thoughts on the subject relate to the exchange of cells between the twins. Blood cells, blood forming cells, and perhaps other cells are exchanged through the fused chorionic blood vessels. Evidence has been presented supporting a theory that X-chromosomes carry genes for both male and female organ development, but that in the absence of Y-chromosomes the genes for male development are suppressed. When Y-chromosomes are present the inhibition of the genes for male development is removed and the genes for female development are suppressed. Each twin contains two genetic populations of cells, one representing its own genotype and one like its twin's. The presence of the Y-chromosome may interfere with or suppress to varying degrees the genes for female development.

The male member of heterosexual twins has generally been thought to be normal in every respect. Based on either theory described here, one would suspect this to be true. There is increasing evidence however that such male twins may also experience reduced reproductive efficiency.

The freemartin condition can be diagnosed in newborn calves by inserting a suitable instrument into the vagina. A sterile plastic inseminating catheter may be used if proper precautions are taken. The vagina of a normal heifer is 12 to 15 cm long while the vagina of the freemartin is only 5 or 6 cm long. Care should be taken not to allow the catheter to enter the suburethral diverticulum in a normal heifer, thus indicating a freemartin. Some individuals prefer to use a 16 mm by 125 mm test tube for this test. In a normal heifer the entire test tube can be inserted into the vagina.

Single birth freemartinism has been reported also. It may result from loss of one embryo from a double ovulation and fertilization after differentiation of sex organs.

**23-2
Infantile
reproductive
system**

An infantile or underdeveloped reproductive system is not usually a permanent cause of reproductive failure (Figure 23-2). It is most frequently found in underfed heifers as characterized by small inactive ovaries, the absence of estrus, and lack of growth of the uterus and vagina. The problem can be corrected by increasing the energy intake, such as turning out to lush spring grazing or by liberal feeding of a good concentrate mixture containing 16% protein. Hormone treatments such as PMSG may be effective, but improved nutrition would be preferred.

Genetically related infantile ovaries that cannot be corrected by hormone treatment or nutrition have been reported. Fortunately the frequency of this condition is low.

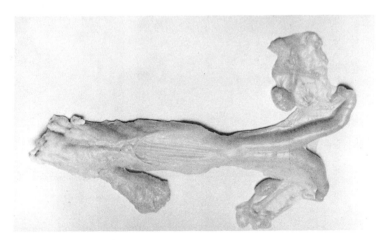

FIGURE 23-2.   Infantile reproductive tract from two year old heifer. Note small uterine horns and small inactive ovaries.

**23-3
Incomplete
structures—
oviduct,
uterus, cervix,
or vagina**

Occasionally females have incomplete oviducts, malformed uteri, blind cervixes, blind vaginas, and the like (Figure 23-3). Usually these defects cannot be repaired, and are difficult to detect without slaughter of the female. These animals come into estrus regularly because the ovaries are functional, but are sterile because the defects prevent sperm and egg from meeting. Some of these defects are inherited and would be expected to occur more frequently if inbreeding was practiced. In fact, some defects are never suspected until they are uncovered by inbreeding. In a normal outcross population the frequency of these structural abnormalities is not great. Even in subfertile dairy heifers, the frequency of incomplete structures was only 3.9%.

Bulls may have similar abnormalities. Most frequently they take the form of incomplete union of the testes with the ducts that transport sper-

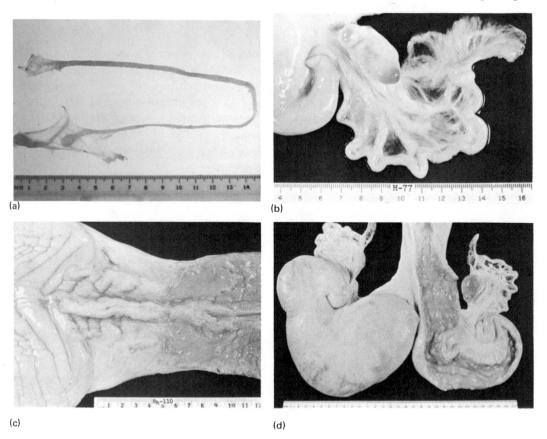

(a)

(b)

(c)

(d)

FIGURE 23-3. Structural abnormalities: (a) an incomplete oviduct; (b) a nodular fibrotic occlusion of the oviduct near the infundibulum; (c) a double cervix; (d) a segmented cornual aplasia. (From Asdell and Bearden. 1959. *Cornell Ext. Bull.* 737; and Tanabe and Almquist. 1967. *Pa. State Univ. Bull.* 736.)

matozoa or of malformations of the penis. The latter condition may prevent copulation.

**23-4**
**Hermaphrodite**

The phenomenon described by the term hermaphrodite is a situation in which the sexuality of an individual is confused by the presence of anatomical structures of both sexes. Hermaphrodites are classified as either true hermaphrodites or pseudohermaphrodites. A true hermaphrodite has both male and female gonads. These may be either separate or combined as ovo-testes. The pseudohermaphrodite has either testes or ovaries, but the remainder of the reproductive system will have parts representing both sexes. Pseudohermaphrodites are usually classified male or female according to the type of gonads present. The male pseudohermaphrodite would have testes but otherwise may be largely female in make-up (Figure 23-4).

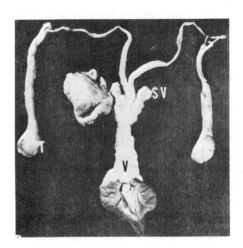

FIGURE 23-4. Internal genitalia of a male bovine pseudohermaphrodite. (From Hafez. 1974. *Reproduction in Farm Animals.* (3rd ed.) Lea and Febiger.)

Genetic sex is determined by the pairing of sex chromosomes. Normal females have an XX pairing; normal males an XY pairing. Deviations can result from abnormalities of either or both of the sex chromosomes, by increases (XXY), or decreases (XO) in the number of sex chromosomes, or by a chimerism of male (XY) and female (XX) cells in the same individual. All of these abnormal sex chromosome combinations have been reported in farm animals. True hermaphroditism with varying combinations of ovaries, testes, and ovo-testes have been described in the horse, goat, cow, and pig. They occur much more frequently in goats and pigs than in the other species. Pseudohermaphrodites are more common than true hermaphrodites.

In the case of pseudohermaphrodites genetic and gonadal sex develop sequentially but there are discrepancies in the development of structures arising from the Wolffian or Mullerian ducts, the urogenital sinus, and external genitalia. In the normal male the development of these structures appears to be dependent on androgen or other substances produced by the gonad. In the male hermaphrodite either the gonads do not produce these substances or the target organs fail to respond to them. In the human female pseudohermaphrodite a deficiency of an enzyme in the metabolic pathway in the adrenal gland causes cortisone not to be produced. This is accompanied by an increase in the production of androgens by the adrenal gland. This syndrome has not been well documented in livestock.

**23-5
Cryptorchid**

Cryptorchidism is a condition in which the testes fail to pass through the inguinal canal into the scrotum. It may be unilateral (one testis remaining in the abdomen and one passing into the scrotum) or bilateral (both testes remaining in the abdomen). The bilateral cryptorchid is sterile since spermatogenesis in mammals cannot occur normally at body temperature. The germinal epithelium of the seminiferous tubules in these males is unaffected until puberty, but if the condition is not corrected by surgery the spermatogonia degenerate and the male will be permanently sterile. The unilateral cryptorchid produces normal, viable, and fertile spermatozoa from the testis which descended into the scrotum. The testis which remained in the body cavity experiences the same fate as the testes of the bilateral cryptorchid.

The interstitial cells of the cryptorchid testes are not affected by the higher temperature. These males, even though sterile, develop normal secondary sex characteristics and normal sexual behavior. The testes of the cryptorchid are smaller than normal for the species. This is presumably due to the failure of the seminiferous tubules to develop normally or to the degeneration of the germinal epithelium after puberty.

Cryptorchidism occurs in most mammalian species. It occurs in man and most domestic animals, but is more common in stallions, goats, and boars than in bulls. The cause of the phenomenon is not fully understood. It may be related to hormonal deficiencies in some cases. In other cases, anatomical obstruction of the inguinal canal or shortness of the ligaments that connect the testes to the body wall may cause the condition. Androgens administered to the foal shortly after birth have resulted in descent of the testes to the scrotum. The other causes require surgery for correction. Whatever the causes, most agree that they are genetically related and therefore the cryptorchid condition occurring in any livestock species should not be corrected by surgery or hormore treatment. Likewise, unilateral cryptorchids should not be used for breeding since both unilateral and bilateral offspring are likely. All cryptorchids, both unilateral and bilateral, should be destroyed with the possible exception of horses, and these should not be used for breeding. Cryptorchid pigs should be destroyed as soon as discovered. If they are reared for meat they will have the characteristic of boars and produce undesirable meat. Cryptorchid bulls and rams may be reared for meat purposes only.

**23-6**
**Injuries**

Injury to the reproductive organs is the cause of significant reproductive failure in both males and females. Some injuries occur spontaneously, but a greater percentage result from manmade causes.

The most common spontaneous injuries to the female reproductive tract occur during parturition. The fetus becomes hyperactive during parturition due to varying degrees of anoxia (Chapter 9). The thrashing of the feet of the fetus has been known to tear the uterus (Figure 23-5). Such injuries may heal without residual effects but may also contribute to breeding failure. Injuries to the vagina may occur during the mating act when males are too vigorous or the penis too large for the female. This is especially a problem with thoroughbred horses.

The most common, and by far the most serious, injuries to the female reproductive tract are manmade. These injuries are primarily inflicted on the cow but some may also occur in the mare. Perhaps the most common relate to adhesions which involve the ovaries and upper portion of the oviducts particularly the infundibulum (Figure 23-6). They are caused by malpractice of the livestock manager or his veterinarian. The injury occurs most commonly during the process of expressing the corpus luteum from the ovary. This is done by placing the end of the thumb against the ovary at the base of the corpus luteum and applying sufficient pressure to expel the corpus luteum from the ovary. This has been a favorite treatment by many individuals for the socalled "retained corpus luteum." In reality a retained corpus luteum may not exist (Chapter 24). In addition to the damage which might be done to the ovary while expressing a corpus luteum, it is impossible to determine whether the infundibulum or the oviduct is between the thumb and the ovary when the pressure is applied. If the serosa of these delicate structures is broken the injured area will grow to any surface that comes in contact with it during the healing process. Thus many oviducts are permanently blocked and infundibula are injured to the point that they may be unable to function normally in capturing the ova as they are released from

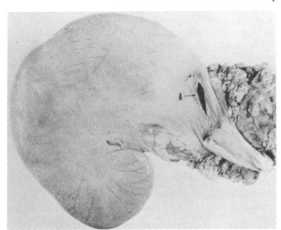

FIGURE 23-5.   Uterus that was torn during the act of parturition. Note point A. (From Asdell and Bearden. 1959. *Cornel Ext. Bull.* 737.)

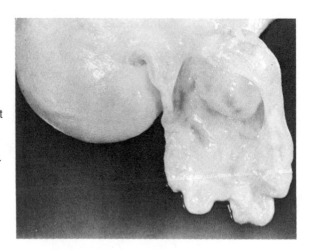

FIGURE 23-6. Obstructive ovarian and infundibular adhesions: partial envelopment of ovary and total incorporation and incapacitation of infundibulum. (From Tanabe and Almquist. 1967. *Pa. State Univ. Bull.* 736.)

the ovary. A corpus luteum, retained or not, should never be expressed from an ovary. In one group of 180 subfertile dairy heifers 6.1% had been injured sufficiently to obstruct conception. An additional 26.1% showed injuries to the ovaries, infundibula, and oviducts that were not sufficiently serious to obstruct conception. Adhesions may also occur as a result of injury caused by manual rupture of follicular cysts in the same manner as described for CL expression.

Other injuries to the reproductive tract usually occur during or immediately following parturition. Too much force and the wrong kind used to pull the fetus may result in injury to the uterus, cervix, vagina, or vulva. A high percentage of cows experiencing dystocia also experience breeding problems and many become permanently sterile. When a female must be given assistance during parturition and the fetus cannot be pulled with relative ease, the livestock manager is advised to enlist the services of his veterinarian unless he is adequately qualified through experience and training.

Injuries to the uterine mucosa are frequently inflicted during manual removal of a retained placenta. Any attempt to disconnect the chorionic villi from the caruncles in the cow's uterus will result in injury. Some individuals unfamiliar with the anatomy of the cow's uterus have been known to tear one or more caruncles loose from the uterus. In view of the possible injury that may occur, the manual removal of the placenta is ill advised. The student is referred to Chapter 19 for management practices related to retained placentae.

Injuries to the reproductive system of the male may also occur. Perforating wounds to the scrotum or the sheath may in themselves cause inflammation or may lead to the introduction of harmful microorganisms that can produce inflammatory conditions. Injuries to the penis have occurred during natural mating and during collection of semen for use in artificial insemination. The most common is the "broken penis." This usually occurs when the penis is bent at a sharp angle during the ejaculatory thrust resulting in a rupture of the corpus cavernosum penis. These males may retain their ability for erection but due to the ruptured area, the distal portion of the penis cannot be raised to horizontal position (Figure 23-7). Venous outlets develop during healing which allows the pressure to bleed off, preventing erection in some cases. Such males may be collected by use of the electroejaculator. A few bulls have had their penuses amputated by tight rubber bands lost from the artificial vagina during collection. These accidents have occurred infrequently, but anyone who collects semen with the artificial vagina is advised to count the rubber bands used both before and after collection.

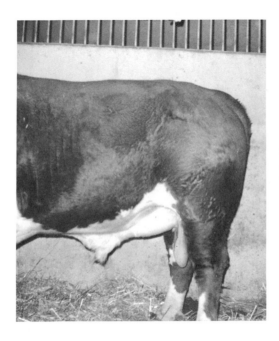

FIGURE 23-7. Bull with a broken penis. Note swelling between sheath opening and testes. (Courtesy of Dr. Leslie Johnson, College of Veterinary Medicine, Oklahoma State University.)

**Suggested reading**

Asdell, S. A. 1955. *Cattle Fertility and Sterility.* Little, Brown and Co. Boston.

Faulkner, L. C. and E. J. Carroll. 1974. Reproductive failure in males. *Reproduction in Farm Animals.* (3rd ed.) ed. E. S. E. Hafez. Lea and Febiger.

Hafez, E. S. E. and M. R. Jainudeen. 1974. Reproductive failure in females. *Reproduction in Farm Animals.* (3rd ed.) ed. E. S. E. Hafez. Lea and Febiger.

Milne, F. J. 1954. Penile and preputial problems in the bull. *J. Amer. Vet. Med. Assn.,* 124:6.

Olds, D. 1978. Inherited, anatomical and pathological causes of lowered reproductive efficiency. *Physiology of Reproduction and Artificial Insemination of Cattle.* (2nd ed.) ed. G. W. Salisbury, N. L. VanDemark and J. R. Lodge. W. H. Freeman and Co.

Roberts, S. L. 1971. *Veterinary Obstetrics and Genital Diseases.* Published by the author. Ithaca, N.Y.

Tanabe, T. Y. and J. O. Almquist. 1967. The nature of subfertility in the dairy heifer. III. Gross genital abnormalities. *Pa. State Univ. Bull.* 736.

# Physiological and psychological causes for reproductive failure

Normal reproduction is a complex series of events which must occur at precisely the right time. The timing mechanism is almost totally dependent on the endocrine system. Disturbances of the endocrine system may induce either temporary or permanent sterility. The disturbance may be the breakage of a link in the total chain and may be temporary or permanent. The total system may be influenced by age, environmental factors, or other factors. The more recently developed techniques for studying the endocrine system have been used primarily to develop a better understanding of normal functions. The application of these techniques to the abnormal situation should make them better understood and provide clues for more effective treatment.

**24-1
Cystic ovaries**

Cystic ovaries are probably among the more perplexing causes of reproductive failure. No treatment can be prescribed that will be effective in all cases. In fact spontaneous recovery has been almost equal to the effect of treatment in some experiments. This is probably related to the fact that different types of cysts occur in the ovaries. Ovulatory failure may result in a thin wall cyst (follicular cyst) or a thick wall cyst (luteinized follicle) in which layers of luteal tissue cover the follicular membrane. A third type of cyst is the cystic corpus luteum. Ovulatory failure without the development of cysts has been reported for all farm species but is much more common in horses and swine than in cattle and sheep. The follicles become partially luteinized and then regress during the estrous cycle in much the same manner as a normal corpus luteum.

**24-1.1
Follicular cysts**

Follicular cysts are more common than other types of cysts. Possibly the ease of detecting accounts for some of the reported higher incidence. These cysts remain thin walled and produce primarily estrogen (Figure 24-1). The result is usually continuous or chronic estrus (3 to 10 days between heats) resulting in the chronic buller or nymphomaniac condition. If the condition continues for an extended period of time, the cow develops an apparent high tail head caused by the relaxation of the pelvic ligaments permitting the pelvis to tilt forward, and a masculine neck, and may bellow like a bull. The cause of cystic ovaries is not known. However, some evidence indicates a deficiency of luteinizing hormone.

Early diagnosis and treatment enhance success. LH, human chorionic gonadotrophin (HCG), or GnRH administered intravenously have been used. GnRH has produced an LH surge similar to a preovulatory surge and initiated ovulation followed by normal estrous cycles in cows with cysts. Progesterone given subcutaneously daily for 15 days has also been used successfully. The cyst regresses and presumably the progesterone causes a buildup of LH in the pituitary. When treatment is terminated and a new follicle develops, sufficient LH is released to cause ovulation and CL formation. All treatments are about 50% effective. Cows that respond to treatment are more likely to become cystic again than cows that never had cysts. Cysts may reoccur after a normal cycle or after a normal gestation. Cows bred at the estrous period following treatment are more likely to produce multiple births.

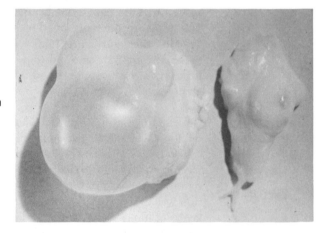

FIGURE 24-1. Ovary with a large thin wall follicular cyst on left. Normal ovary on right. (From Asdell and Bearden. 1959. *Cornell Ext. Bull.* 737.)

GnRH, a small peptide molecule, is less antigenic than LH or HCG which are large protein molecules. Cows are less likely to produce antibodies against GnRH and it remains effective on repeated use. LH and HCG may lose effectiveness by second or third treatment if the cow continues to be a problem.

Follicular cysts are most common in dairy cattle and in swine but rarely occur in beef cattle, sheep, and horses. In swine, estrous cycles may be irregular or prolonged, but nymphomania does not occur.

**24-1.2
Luteinized
follicles**

A high percentage of thin wall cysts become luteinized. The luteal tissue produces progesterone and the follicular fluid is usually high in progesterone. Multiple cysts are common, resulting in a greatly enlarged ovary (Figure 24-2). The condition is usually characterized by long periods of anestrus. LH is not effective as a treatement due to the thickness of the follicular wall. Oral progestins have been used and are usually effective in causing the regression of the luteinized follicles. The material must be fed for 2 to 3 weeks or until the cyst has regressed as determined by rectal palpation. PGF$_2\alpha$ may prove to be an effective treatment.

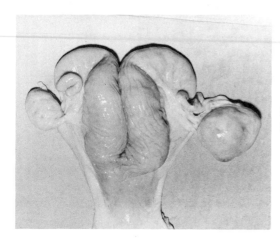

FIGURE 24-2. Ovary on right has multiple large luteinized follicles. Ovary on left is normal.

**24-1.3
Cystic corpus
luteum**

The cystic CL differs from the luteinized follicle even though some similarities exist. While the cystic CL contains a cavity which is usually filled with fluid, its origin is the same as that of a normal corpus luteum. In the early stages of development of the CL the theca interna of the follicular membrane proliferates rapidly and folds into the crater left by the ruptured follicle, thus creating a small cavity. Some reports indicate that 25% of these cavities are never completely filled with granulosa cells growing from beneath and a small cavity persists throughout the life of the CL. The question arises as to which of the cavities are pathological and which are normal. Some workers have set arbitrary maximum sizes for normal cavities, usually 7 or 8 mm in diameter. Larger cavities are considered cysts and pathological (Figure 24-3). The cystic corpus luteum is considered to contribute to early embryonic mortality. In one large study no pregnancies were observed with cysts larger than 7 mm in diameter. In another study of aged infertile cows a high percentage had cystic CL, whereas none were seen in aged fertile cows. Nonpregnant animals with cystic corpora lutea exhibit essentially normal cycles, but abnormal estrous periods have been reported. Based on physiological evidence it has been postulated that three types of ovulatory dysfunction—delayed ovulation, follicular cysts, cystic CL—stem from the same basic inadequacy of luteinizing hormone but reflect different degrees of insufficiency.

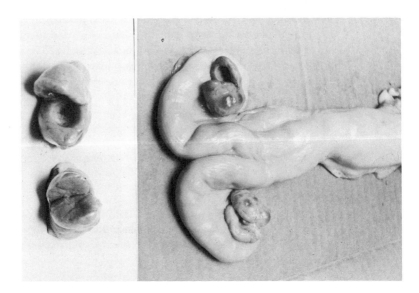

FIGURE 24-3. The right ovary (top) contains a cystic corpus luteum. A normal corpus luteum is shown at bottom left.

**24-2
Retained
corpus luteum**

Much has been said and written concerning the retained corpus luteum. Many cows have been treated and mistreated because of this elusive phenomenon. In the authors' opinion the corpus luteum probably is retained only in association with some pathological condition of the uterus. Pyometra, mummified fetus, or other conditions which cause the uterus to react as a pregnant uterus cause the retention of the corpus luteum resulting in suppression of the estrous cycle. The persistence of the corpus luteum in these cases is due to an interference of the uterine luteolytic mechanism (release of PGF$_2\alpha$, Chapter 5). The corpus luteum of the preceding pregnancy is probably never retained since luteolysis is a normal part of parturition initiation (Chapter 9).

Many cows have been diagnosed as having retained CL that were simply between periods of estrus. Either these cows ovulated without overt symptoms of estrus or estrus was undetected. It is impossible for a veterinarian or a livestock manager to differentiate between a normal CL and the "retained CL" by rectal palpation. The only way to be sure that the CL is not normal would be by repeated palpations at 3- or 4-day intervals for a period of 2 weeks. If the cow does not return to estrus and the CL does not regress then a retained CL would exist. For practical purposes if the uterus is found to be empty (not pregnant and does not contain a mummified fetus, pus, or other fluids) and a corpus luteum is detected on an ovary it is safe to assume that the animal is cycling.

Some cows exhibit long estrous cycles due to embryonic mortality. If the embryo death occurs by 12 days post-breeding the cow will return to estrus in the usual length of time. When death occurs later than this the return to estrus is delayed depending on when the death occurs. The later the death occurs, the longer the interval between death and return to estrus. Some of these cows have also been classified as having retained CL. In reality these are cases of corpora lutea of pregnancy and do not fit the classical definition of a retained corpus luteum.

**24-3**
**Anestrus**

Anestrus is defined as the absence of estrus. There are several causes of anestrus but the most common one is pregnancy. In cows and mares that can be rectally palpated for pregnancy no female should be treated for anestrus until she is checked for pregnancy. Most cows exhibit a period of anestrus following parturition. In dairy cows this period lasts from 30 to 40 days, but in beef cows that are suckling calves the anestrus period may be greatly extended. The sow usually has an estrous period a few days after parturition but will not cycle again until the piglets are weaned. She usually will cycle 3 or 4 days after the piglets are removed. The mare usually exhibits a foal heat 9 or 10 days post-parturition, and most will cycle again about 30 days post-partum. Most mares are seasonal breeders and are stimulated to sexual activity by increasing daylength in the spring. They will usually cycle through November, and if they do not become pregnant will go into anestrus until the following spring. Most breeds of sheep are also seasonal breeders and are stimulated to sexual activity by decreasing daylengths. Most ewes will exhibit one or more quiet ovulations at the beginning of the breeding season. Some flocks have had their breeding season extended by selection and management to the extent that three lamb crops can be obtained in a two-year period.

The student is referred to section 18-2 for the treatment of anestrus in the cow.

Low-energy intake levels have been shown to delay puberty in all species. This results in anestrus at a time when these females are old enough for the first breeding. Low energy levels have also caused anestrus and particularly have contributed to prolonged anestrus following parturition in beef cattle. This topic is more thoroughly discussed in Chapter 22. About 3% of the cows in dairy herds where nutrition should not be a factor have been found to be in anestrus. Various treatments including FSH and estrogen have been used.

**24-4**
**Irregular**
**estrous cycles**

The normal estrous cycle length in the cow has been accepted as 18 to 25 days. Similar variations for the other species would also be expected as a normal range. The data in Table 24-1 show the distribution of estrous cycle lengths for 200 cows through 500 cycles. The ovaries of these cows were palpated at least once weekly so that accurate determinations were made relative to cycle length and ovarian involvement. It is interesting that only 60.4% of the cows had normal length cycles. An additional 29% had either short (1.4%) or long (27.6%) cycles. Most of the remaining cows (9.8%) ovulated without demonstrating estrus previous to the first postpartum estrus. Four cows (0.8%) demonstrated estrus but either failed to develop a follicle and/or failed to develop a corpus luteum. Of the 27.6% that had long estrous cycles, 8.8% were long because of "silent heats" following a normal estrous cycle. An additional 3.6% returned to anestrus following a normal postpartum estrus. The remaining 15.2% were reported as having retained corpora lutea. The researchers did not report the condition of the uterus in cows diagnosed to have retained corpora lutea, nor did they mention the probability of embryonic mortality as a cause of delayed return.

TABLE 24-1. Number and percentage of normal and abnormal intervals between heats, as determined from ovarian examination of 200 cows through 500 cycles

| Condition | Intervals | | |
|---|---|---|---|
| | Number | % | |
| Normal estrous cycles—18-25 days | 302 | 60.4 | |
| Estrous cycles <18 days—not cystic | 7 | 1.4 | |
| False estrus <18 days | 4 | .8 | |
| Silent estrus previous to first postpartum estrus | 49 | 9.8 | |
| Estrous cycles >25 days | 138 | 27.6 | |
|    Silent estrus after the first postpartum estrus | | 44 | 8.8 |
|    Persistent corpus luteum | | 76 | 15.2 |
|    Anestrus—smooth ovaries | | 18 | 3.6 |
| Total | 500 | 100.0 | |

Adapted from Trimberger and Fincher, Cornell Exp. Sta. Bul. 911, 1956.

Other researchers have reported from 16% to 44% of the estrous cycles occurring shorter or longer than 17 to 24 days. The higher figures were reported for large-scale studies using nonreturns to artificial insemination. These studies were also made during a time when the disease vibriosis was being transmitted through AI and a major symptom of this disease is embryonic mortality. A high percentage of cows infected with vibriosis return to estrus between 27 and 55 days post-breeding. Other factors that may have contributed to these high figures are "silent estrus", undiagnosed estrus, and one or more natural services (information not available to the researchers) between the AI services.

The conception rate for cows bred during the period of estrus following short estrous cycles is much lower than when breeding follows a normal length cycle. Conception is also lower following cystic ovaries. However, when cows are bred at the first estrus following long cycles, "silent estrus" and anestrus the conception rate was equal to that attained for cows bred following normal cycles as shown by the data in Table 24-2. This could be expected since short cycles and cystic ovaries are indicative of endocrine imbalances. The periods of estrus following "silent heats" and anestrus were normal from an endocrine standpoint. The same can be said for those periods of estrus following a retained corpus luteum, since there is a good possibility that the delayed regression of the corpus luteum was related to embryonic mortality rather than endocrine imbalance.

TABLE 24-2. Conception rates in cows bred during and following various ovarian conditions

| Estrus and ovarian conditions | Cows bred | Conception to one service |
|---|---|---|
| Cows bred in control group* | 200 | 60.0 |
| Cows bred in experimental group† | 200 | 65.0 |
| Artificially bred during silent estrus | 20 | 65.0 |
| Normal estrus after silent estrus | 48 | 62.5 |
| Estrus after persistent corpus luteum | 46 | 65.2 |
| Estrus after anestrus—smooth ovaries | 15 | 80.0 |
| Service after recovery from cystic ovaries | 34 | 41.2 |
| Service following short cycles—<18 days | 36 | 33.3 |

Adapted from Trimberger and Fincher, Cornell Exp. Sta. Bul. 911, 1956.
*Not palpated, bred at observed estrus.
†Palpated regularly to determine condition of ovaries.

**24-5
"Silent estrus"**

This term is defined as ovulation without behavioral manifestation of estrus. The data in Table 24-1 show that 9.8% of the cows had a silent estrus preceding the first observable estrus following parturition. An additional 8.8% had a silent estrus after the first observable estrus following parturition. Of the 500 estrous cycles detected by palpation 18.6% involved silent estrus. The cows in this study were observed twice daily with other cows in an exercise yard. In addition, they would not stand to be mounted by "indicator" cows (cows treated with diethylstilbestrol) or by a teaser bull when checked at 2- to 4-hour intervals. They were also tried for natural breeding without success. The only sign of estrus was the characteristic mucous discharge and the presence of a follicle on the ovary. Other reports have shown the incidence of silent estrus to be as high as 27.3%. Twenty cows were artificially inseminated during silent estrus during which they would not stand for natural mating. Of these 65% conceived. Inseminations were timed by frequent palpations of the ovary to determine when the follicle was reaching maturity. The follicle bulges beyond the contour of the surface and is more turgid during the last several hours before ovulation than it is previously.

No treatment or corrective measure is known for silent estrus. The physiological cause is not known, but subthreshold hormone production or balance between hormones may be factors. The intervals preceding and following silent estrus are usually normal. Since ovulation occurs followed by a normal corpus luteum, no disturbance from a functional standpoint is indicated. Usually the estrus following a silent estrus will be indicated by normal symptoms but occasionally two or three silent estrous periods occur in succession.

**24-6**
**Age**

The effect of age on reproductive efficiency is difficult to measure in farm animals. The selection pressures for both producing ability and reproductive ability eliminate many animals from the herds and flocks at relatively early ages. Additional environmental factors that affect reproductive efficiency also tend to mask the effects of age. A number of studies in cattle have shown that heifers have a lower reproductive efficiency than cows. It is true that most anatomical abnormalities that interfere with normal reproduction are discovered in heifers and eliminated. When these are eliminated the remaining heifers under excellent management conditions should have a higher reproductive efficiency than cows. Unfortunately the managers of many herds do not spend adequate time checking heifers for estrus so that inseminations can be properly timed. Many heifers will show symptoms of estrus for 6 to 12 hours before standing when mounted. Cows will usually only show estrual symptoms for 2 to 4 hours before standing. These factors probably contribute to the lowered reproductive efficiency reported for heifers.

The data in Table 24-3 indicate that reproductive efficiency is greatest for dairy cows 3 to 4 years of age. Conception began decreasing in the 5- to 7-year age group and a marked decrease in reproductive efficiency occurred in cows over seven years of age. Studies with beef cattle show considerably more variation but generally the highest level of reproductive efficiency is seen between four and nine years of age. Similar effects of age in the other species can be expected.

Reproductive efficiency in the male is also affected by advancing age. Studies have shown that the fertility of bulls used in artificial insemination reaches a peak between 2 and 4 years of age. The fertility peak may occur prior to the peak in sperm production. The percentage of motile sperm after freezing and thawing and the percentage of morphologically normal sperm at the time of collection are lower for older bulls than for younger ones. Even though these statements are generally true, there is considerable individual variation. There are many older bulls being used in artificial insemination that produce semen with fertilizing capabilities as high as the younger bulls.

The cause of lowered reproductive efficiency with advancing age is not known. It may be due to hormonal imbalance or a deficiency, which contributes to reduced ovulation rate or abnormal spermatogenesis. It may also be related to a deterioration of the gametes which in turn may affect fertilization rate or contribute to an increased embryonic mortality rate.

TABLE 24-3.   Percent conception for cows by age groups compared with cows <36 months of age

| Age | 1st service | 2nd service | 3rd service |
|-----|-------------|-------------|-------------|
| Mo. | average percentage unit | | |
| 36-47 | 1.0 | .9 | 1.9 |
| 48-59 | −1.9 | −3.8 | 1.2 |
| 60-84 | −6.2 | −3.1 | −6.4 |
| >84 | −14.8 | −12.7 | −6.9 |

Adapted from Spalding et al., J. Dairy Sci. 58:718, 1975.

**24-7**
**Seasons of the year**

Cattle and swine are considered to be continuous breeders while horses and sheep tend to be somewhat seasonal in their breeding patterns. Even with cattle and swine there are differnces in the reproductive efficiency due to season.

**24-7.1**
**Cattle**

Many studies have been conducted with both cows and bulls relative to the effect of season on reproductive efficiency. Essentially all studies show that reproductive efficiency for cows is greatest during the spring. This is probably due to the flushing effect from spring grass. Light may be a factor also. A slightly lower number of studies have shown that reproductive efficiency is lowest during the summer, and a few studies conducted in the extreme northern part of the United States and in Canada have shown that reproductive efficiency is lowest during the winter. All of the studies in the southern half of the United States have shown that reproductive efficiency is lowest during the summer, with little difference between fall and winter, those two seasons falling about halfway between spring and summer. The stress caused by high temperature and high humidity are accepted as the cause of lowered reproductive efficiency during the summer months (Chapter 21). Short daylength has been indicated as the cause of lowered reproductive efficiency in the northern area but management practices contribute to the problem also. Traditionally cattle were housed in stanchion barns during the winter months and many still are. Diagnosing estrus under these conditions is much more difficult and contributes to lowered reproductive efficiency. See Table 19-4.

The effects of season on the fertility of semen produced by bulls is very similar to the seasonal effects on cows. The lowest quality semen is produced during August and September. Some studies have shown a gradual decline in semen quality as early as late spring. Other studies have shown as with cows that short daylengths in the extreme north have a greater adverse influence on semen quality than the mild summers of the same region.

The physiological relationship of environmental stress to reproduction has been covered in Chapter 21 and will not be repeated here.

**24-7.2**
**Swine**

Slight variations in the fertility of sows have been attributed to season. Sows artificially inseminated during the summer had a lower farrowing rate (52%) than in other months (62%), and litter size was smaller (9.92 as compared to 10.46). Boars have also been shown to produce lower quality semen during the summer than during other seasons of the year. Ejaculates from boars are lower in volume and higher in concentration during cold temperatures compared to warmer temperatures. Managers have reported lower conception rates and smaller litter size during extremely cold winters compared to the usual winter cold. This effect of extreme cold may be due to general stress mechanisms which were discussed in Chapter 21.

**24-7.3**
**Sheep**

Since sheep are seasonal breeders and are stimulated to sexual activity during long daylengths, the breeding season usually begins in July. Rams, like males of other species, produce lower quality semen during late summer particularly in the warmer regions. Delaying the breeding seasons until late September improves reproductive efficiency but management systems and lamb marketing must be taken into consideration.

**24-7.4**
**Horses**

Mares are seasonal in their breeding behavior also. They are stimulated to sexual activity by increasing daylength which results in the peak breeding season occurring during the spring. Since this is the season when reproductive efficiency generally is highest, there is less adverse effect from season in horse reproduction.

**24-8**
**Psychological**
**disturbances**

The relationship between psychological phenomena and reproduction has not been researched throughly and is poorly understood. Observations by alert managers, data from a few experiments designed to provide psychological data, and data gleaned from unrelated research have led to the general conclusion that handling animals gently aids both production and reproduction.

**24-8.1**
**Females**

Two extensive studies involving both dairy and beef cattle have shown that cows that were extremely nervous or excited at the time of insemination had a lower conception rate. The degree of nervousness may have been an indication of the amount of adrenalin released. Adrenalin is known to interfere with contractions of muscles normally stimulated to contractions by oxytocin. Contractions of the muscles of the uterus and oviduct are involved in the transport of sperm to the site of fertilization. Therefore, it is conceivable that the adrenalin released by these nervous cows may have interfered with sperm transport. Release of ACTH and glucocorticoids which suppress LH may also be a factor.

Both dairy and beef cows that were isolated from the remainder of the herd at the beginning of estrus and confined until the time of insemination showed lower conception rates than those cows that were left with the rest of the herd until time for insemination. Conception rate was affected more in the beef cows than in the dairy cows. The confinement apparently caused stress which resulted in lower conception rate. Confinement of the cows after insemination lowered conception rate in beef cows, but dairy cows were not affected. The probable reason that dairy cows were less affected when confined both before and after insemination is that dairy cows are more accustomed to being handled and confined than are beef cows. Driving cows from the pasture and separating them from the rest of the herd at time of insemination had no effect on conception. This management practice appears to create less trauma than confining the estrous cow before or after insemination.

Cows that are difficult to inseminate have shown lower conception than those in which insemination was accomplished more easily. This may have been caused by extra excitement or trauma on the part of the cow. Admittedly, it may also have been caused by the inseminators not properly placing the semen. Lower nonreturn rates are reported for inexperienced AI technicians for the first 3 months (Table 24-4). Part of this may be related to the extra time required to pass the catheter through the cervix and the resulting excitement and trauma.

The sow's reproductive performance is also affected by psychological

TABLE 24-4. Breeding efficiency of inexperienced inseminators in areas where AI was new and in areas where AI had previously been established

| Mo. of operation | New areas | | Old areas | |
|:---:|:---:|:---:|:---:|:---:|
| | No. areas | % 60-day NR | No. areas | % 60-day NR |
| 1st | 29 | 49.0 | 14 | 57.0 |
| 2nd | 26 | 52.0 | 14 | 62.8 |
| 3rd | 24 | 61.0 | 12 | 64.8 |
| 4th | 22 | 64.5 | 12 | 65.4 |

From Olds and Seath, Ky. Agr. Exp. Sta. Bul. 605, 1954.

factors. The overt signs of estrus are more clearly demonstrated when boars are within hearing and smelling range. Recordings of boars' mating noises or pheromones isolated from the boar have been shown to be stimulating in the absence of boars. The effect of these factors on conception and litter size have not been reported.

Gilts reared in confinement have suppressed estrus as long as they remain in the same pen. Moving them to an adjacent pen has some stimulating effect, but moving them to a pasture area has a greater effect. Many gilts will come into estrus within 2 to 3 days after being moved. Even gilts reared on pasture are stimulated by movement to new quarters. Gilts have shown response from being moved out of a pen for a few minutes and put back into the same pen.

Gilts not moved to a different pen or location will fight a boar that is put in with them for mating. If the boar is of similar age and size he may be seriously injured or even killed. The boar can be introduced at the same time the gilts are moved to a new location and little animosity is shown by the gilts.

Ewes show no overt symptoms of estrus in the absence of the ram even though ovulation occurs. Ewes appear to be stimulated to start cycling earlier at the beginning of the breeding season when rams are running with them.

## 24-8.2
## Males

Bulls in AI centers have demonstrated several psychological effects on performance. Certain bulls develop inhibitions about certain teaser animals or a particular collection area. These can be overcome by changing the teaser animal and/or collecting the bull at a different location.

Pain or discomfort related to copulation or AI collection has caused loss of libido in bulls. An artificial vagina with too high a temperature will cause a bull to be reluctant to mount for ejaculation. If the AV is not held at the proper angle, the resulting pain may have similar results. No data seem to be available on the effect of mistreatment prior to or at the time of collection on quality or quantity of semen produced.

Simulated mistreatment by daily injections of adrenalin for 10 weeks has shown a marked decrease in sperm output by bulls compared to control bulls. Daily shocking of bulls with an electric cattle prod did not reduce sperm production. However, any stress or change in routine causes an increase in abnormal sperm in some bulls. All types of abnormalities increase but cytoplasmic droplets appear first and are last to disappear.

Based on what is known for both males and females, animals should be handled in a manner that produces the least amount of stress possible. Treatment resulting in undue excitement or pain should be avoided. Both reproduction and production will be enhanced.

**Suggested reading**

Bishop, M. W. H. 1970. Aging and reproduction in males. *J. Reprod. Fert. Supp.*, 12:65.

Carpenter, L. M. 1976. A study of management factors as they relate to conception rate in cattle artificially inseminated. PhD. Dissertation, Miss. State Univ.

Gilmore, L. O. 1949. The inheritance of functional causes of reproductive inefficiency: a review. *J. Dairy Sci.*, 32:71.

McEntee, K. 1958. Cystic corpora lutea in cattle. *Internat. J. Fert.* 3:120.

Olds, D. and D. M. Seath. 1954. Factors affecting reproductive efficiency in dairy cattle. *Ky. Agr. Exp. Sta. Bull.* 605.

Roberts, S. J. 1971. *Veterinary Obstetrics and Genital Diseases.* Published by the author. Ithaca, N. Y.

Salisbury, G. W., N. L. VanDemark and J. R. Lodge. 1978. *Physiology of Reproduction and Artificial Insemination of Cattle.* (2nd ed.) W. H. Freeman and Co.

Spalding, R. W., R. W. Everett and R. H. Foote. 1975. Fertility in New York artificially inseminated Holstein herds in dairy herd improvement. *J. Dairy Sci.*, 58:718.

Tanabe, T. Y. and J. O. Almquist. 1967. The nature of subfertility in the dairy heifer. III. Gross genital abnormalities. *Pa. State Univ. Bull.* 736.

Trimberger, G. W. and M. G. Fincher. 1956. Regularity of estrus, ovarian function, and conception rates in dairy cattle. *Cornell Exp. Sta. Bull.* 911.

Whitmore, H. L., W. J. Tyler and L. E. Casida. 1974. Incidence of cystic ovaries in Holstein-Friesian cows. *J. Amer. Vet. Med. Assoc.*, 165:693.

## Chapter

# 25 Infectious diseases that cause reproductive failure

The material presented in this chapter will deal with those diseases that contribute to reduced reproductive efficiency, cause abortions, or adversely affect the offspring which are carried to term. Some are venereal diseases, affecting only the reproductive processes, while others are general systemic diseases with the effects on reproduction being secondary. Some of the diseases have public health significance and this will be alluded to. This presentation will be limited to those diseases that have economic significance to the livestock producer. Some of the diseases may strike suddenly and cause extensive losses in production and reproduction in a very short period of time. Others are more chronic in nature and may not be recognized but cause significant economic loss over a long period of time.

This chapter is not intended to train students in the field of veterinary medicine, but rather to make them aware of the diseases that cause reproductive failure and to acquaint them with symptoms, methods of transmission and methods of control so that they can be better livestock managers, and hopefully avoid many disease problems through these management techniques. Table 25-1 summarizes the diseases that will be covered in this chapter from the standpoint of affected species, causative organisms, method of transmission, major effects on reproduction, and control measures.

**25-1
Bacterial
diseases**
25-1.1
Vibriosis

Vibriosis affects both cattle and sheep. The causative organism for cattle is *Vibrio fetus venerealis*, while *Vibrio fetus intestinalis* causes the disease in sheep (Figure 25-1). The distribution of the disease is worldwide and is found throughout the United States.

**25-1.1a Method of transmission**   *Vibrio fetus venerealis* in cattle is a venereal disease. Transmission other than by natural mating and through infected semen has not been reported. Even when experimentally infected cows are confined with noninfected cows, the disease is apparently not

TABLE 25-1. Summary of diseases affecting reproduction in farm species

| Disease | Affected species | Causative organism(s) | Effects on reproduction | Method of transmission | Control measures |
|---|---|---|---|---|---|
| **Bacterial Diseases** | | | | | |
| Vibriosis | Cattle | Vibrio fetus venerealis | Embryonic mortality Early abortion | Sexual contact Contaminated semen | Vaccination Breed AI |
| | Sheep | Vibrio fetus intestinalis | Abortion last trimester | Contaminated feed and water | Vaccination |
| Leptospirosis | Cattle | L. pomona L. canicola L. grippotyphosa L. icterohemor-rhagiae L. hardjo | Abortion last trimester | Urine contaminated feed water and air, Contaminated semen Wildlife | Annual vaccination Antibiotic therapy of acute cases Sanitation |
| | Swine | All but L. hardjo | Late abortion Weak pigs | Same | Same |
| | Horses | All but L. hardjo | Late abortion | Same | Same |
| Brucellosis | Cattle | Brucella abortus | Abortions Weak calves Retained placentae Reduced breeding efficiency | Contaminated feed and water from aborted material and at calving Contaminated semen | Calfhood vaccination Test and slaughter Prevent exposure |
| | Sheep | Brucella melitensis | Abortion | Same | Vaccination Test and slaughter |
| | Swine | Brucella suis | Abortion Weak pigs | Same | Test and slaughter |

301

TABLE 25-1. continued

| Disease | Affected species | Causative organism(s) | Effects on reproduction | Method of transmission | Control measures |
|---|---|---|---|---|---|
| **Bacterial Diseases** | | | | | |
| *Listeriosis* | Cattle, Sheep | *Listeria mono- cytogenes* | Late abortion Retained placentae Metritis | Contaminated environment | Sanitation Antibiotic therapy |
| *Nonspecific uterine infections* | Cattle | Variety of bacterial organisms | Extended anestrus Reduced breeding efficiency | Contaminated calving area Introduced with treatments etc. | Sanitation |
| *Contagious Equine Metritis* | Horses | Bacterium similar to *Moraxella* | Endometritis, Cer- vicitis, Vaginitis | Natural mating Contaminated equip. | Sanitation Antibiotic therapy |
| **Protozoan Diseases** | | | | | |
| *Bovine Trichomoniasis* | Cattle | *Trichomonas fetus* | Early abortion Pyometra Sterility | Sexual contact | Sexual rest Breed AI Slaughter infected bulls |
| *Toxoplasmosis* | Sheep, Cattle, Swine | *Toxoplasma gondii* | Late abortions Weak young Still births Retained placentae | Ingesting oocysts from contaminated environment | Prevent ingesting oocysts and contaminated carcasses |
| **Viral Diseases** | | | | | |
| *Bovine Viral Diarrhea* | Cattle | BVD virus | Abortion Fetal abnormalities | Contaminated environment Virus comes in contact with mucous membrane Infected semen | Vaccination Booster vaccination may be desirable |

| Disease | Animal | Agent | Effect | Transmission | Control |
|---|---|---|---|---|---|
| *Infectious Bovine Rhinotracheitis* or *Pustular Vulvovaginitis* | Cattle | IBR-IPV virus | Abortion in 2nd half of gestation; Temporary infertility | Contaminated environment; Virus comes in contact with mucous membrane; Infected semen | Vaccination (IM or nasal); Booster vaccination with nasal preparation desirable |
| *Equine Rhinopneumonitis* | Horses | Equine herpesvirus I | Abortion last trimester | Virus comes in contact with mucous membrane by aerosols | Vaccination July and October |
| *Equine Viral Arteritis* | Horses | Equine arteritis virus | Abortion in 2nd half of gestation | Virus comes in contact with mucous membrane by aerosols | Isolation of apparently infected mares during abortion and parturition |
| *Bluetongue* | Sheep, Cattle | Bluetongue virus | Damage to central nervous system of fetus | *Culicoides* gnat, sheep ked; Possibly by infected semen | Vaccination of nonpregnant animals |

**Mycoses**

| Disease | Animal | Agent | Effect | Transmission | Control |
|---|---|---|---|---|---|
| *Mycoses* | Cattle | *Aspergillus fumigatus*; *A. absidia*; *A. mucor* | Abortion in mid and last trimester | Probably inhalation of spores from moldy feeds | Sanitation and care in feed storage |

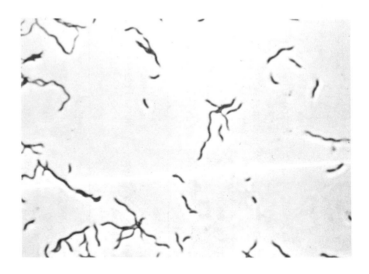

FIGURE 25-1. Photomicrograph of *Vibrio fetus venerealis* organisms. (Courtesy of R. Hidalgo. College of Vet. Med., Miss. State Univ.)

transmitted. Figure 25-2 shows how highly contagious the disease can be when introduced into herds practicing natural mating.

The four herds involved were grade Holstein herds in western New York. Herd A bought four cows from a fifth herd that was infected with vibriosis. In the process of trying to get the cows pregnant, four bulls that were being used in the herd became infected. In turn these bulls transmitted the disease throughout the herd. The owner thought that the problem of low fertility in his herd was due to his bulls, so he borrowed a bull from herd B. After finding that this bull did not settle the cows either he was returned to herd B and eventually infected that entire herd. The owner of herd C felt sorry for the owner of herd B because his bulls were sterile, so he loaned him a young bull that had been very fertile in his own herd. This bull did not settle cows in herd B and was returned and used in herd C. The first 19 cows serviced by this bull after returning to herd C all failed to conceive. Vibriosis was transmitted to other bulls in the herd and eventually to the entire herd. Herd D was a small herd of only 30 cows and one bull. The owner had two calf crops from the bull he was using and did not want to breed his daughters back to him so he arranged to trade his bull for a bull that had been used and infected in herd B. His herd became infected also.

Vibriosis can be transmitted between bulls in AI situations. When the same teaser animals are used for both infected and noninfected bulls, it is difficult to prevent the infected bull from leaving organisms on the teaser animals. If the clean bull touches the same area with his glans penis and

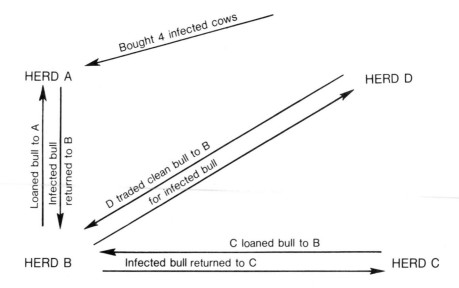

FIGURE 25-2.   Diagram of vibriosis transmission in an actual case study in New York State.

picks up the organism he becomes infected. Fortunately the commerical AI organizations are now maintaining vibriosis-free studs. The *Vibrio fetus venerealis* organism is shed in the semen of infected bulls. If this semen is not properly treated with antibiotics (see Section 16-3), vibriosis will be transmitted to the cows bred. The organism will survive freezing and thawing, so it is essential that semen to be used for artificial insemination comes from bulls free of vibriosis or that it be properly treated with antibiotics to kill the *Vibrio fetus venerealis* organism.

In sheep vibriosis, the *Vibrio fetus intestinalis* organism may be ingested with feed and water contaminated by fluids and discharges from aborting ewes. The infection may persist for some time in the gall bladder of infected animals and the organisms shed may continue to contaminate the environment.

### 25-1.1b Symptoms in cattle

1.   The most common symptom is repeat breedings. The average number of natural matings required for conception is 5 with a range from 2 to 25.

2.   The estrous cycles are long and irregular, ranging from 25 to 55 days in length. These long cycles are caused by early embryonic mortality. The death of the embryo occurs late enough to postpone the return to estrus.

3.   The estrual mucus will be cloudy. This cloudiness is caused by pus, and flakes of pus may be observed in the mucus.

4.   Cows usually develop sufficient immunity after several services to carry a calf to term.

5.   About 10% of the infected cows will have noticeable abortions. Most of these will occur at 2 to 3 months of gestation.

The bull demonstrates no clinical symptoms of the disease, even though the organism may persist in the reproductive tract for an extended period of time. Spontaneous recovery by bulls has been reported.

**25-1.1c Symptoms in sheep**   The effect of vibriosis in sheep is usually much more dramatic than in cattle. In order for the disease to exhibit reproductive symptoms it must occur 4 to 6 weeks prior to lambing. Abortion rates up to 70% have been reported. The disease is fatal to a small percentage of ewes and these show significant necrosis of the caruncles.

**25-1.1d Diagnostic procedures**   The symptoms of vibriosis are not different enough from the symptoms of other diseases in either cattle or sheep to be relied on as a diagnostic procedure. In herds where breeding dates are kept and close observations are made at the time of estrus, one can strongly suspect vibriosis from the symptoms. In beef herds in which the bulls are run with cows, vibriosis should be suspected with prolonged breeding and calving seasons. In either case a positive laboratory diagnosis is desirable to rule out the possibility of other diseases.

Isolation and identification of the organism is the most definitive test used. However, *Vibrio* organisms are very difficult to isolate because they grow slowly. Most samples available for culturing are contaminated with large numbers of fast growing organisms that rapidly overgrow the culture media. Stomach fluid from aborted feti, estrous mucus, and semen from bulls are preferred materials for culture. A diagnosis should not be made based on a single negative culture. Several negative cultures on the same animal or negative cultures on several problem animals make a diagnosis more reliable. Selected antibiotics in the culture media to control contaminants but allow the *Vibrio* organism to grow are helpful.

Antibodies appear in the female reproductive tract. Cervico-vaginal mucus can be tested for the presence of these antibodies. The test is usually considered to be a herd test and samples of mucus should be taken from six to eight problem cows during diestrus about 60 days after first breeding. A positive agglutination test results when *Vibrio* organisms added to the antibody preparation form clumps. A more sensitive procedure is obtained when soluble antigens from the organism are absorbed onto red blood cells or latex particles. These cells or particles agglutinate in a positive test.

Bulls are sometimes tested for vibriosis by breeding virgin heifers negative to vibriosis naturally or with untreated semen and testing the heifers. In some cases where large numbers of bulls are to be tested prepucial scrapings are pooled from a few bulls and deposited in test heifers. In the case of a positive test the individual bulls in the small group would have to be tested.

In sheep the diagnosis may be made by direct microscopic observation or by the culture of the *Vibrio fetus intestinalis* organism from the placenta or from the aborted fetus.

**25-1.1e Control measures**   Prevention is preferable to treatment. Vibriosis is virtually unheard of in dairy herds practicing 100% AI and using semen from commercial semen-producing organizations. If natural mating is to be utilized, maintaining a closed herd has a lot of merit in preventing several diseases including vibriosis. When male or female replacements must be purchased it is much safer to buy prepuberal animals. If postpuberal females are purchased, heifers in late pregnancy are safer than cows. Purchased postpuberal animals should be isolated and adequately tested before they are added to the breeding herd.

Cows infected with vibriosis will develop immunity and/or experience spontaneous recovery from the disease if allowed 60 to 90 days sexual rest. Intrauterine infusion with 0.5 to 1 gm of streptomycin 24 hours after insemination or one estrous period prior to insemination will restore fertility. Bulls have been successfully treated with large subcutaneous doses of streptomycin accompanied by local treatment of the sheath and glans penis with the same material. Because of the elusiveness of the organism it is difficult to know without exhaustive testing when a bull is free of the organism.

Vaccination with a bacterin prepared from respective sub-species is available for both cattle and sheep. Cattle should be vaccinated approximately two months prior to breeding and annual vaccination may be indicated. Sheep may be vaccinated prior to the breeding season or during early pregnancy.

**25-1.2**
**Leptospirosis**

Leptospirosis is not a venereal disease although it does seriously affect the reproductive processes in cattle, horses, and swine. The disease also causes many symptoms totally unrelated to reproduction. There are many serotypes of the genus *Leptospira* which cause leptospirosis in wild animals, domestic animals, and man. Five serotypes have been associated with leptospirosis in farm animals.

**25-1.2a Causative organisms**   The serotypes responsible for leptospirosis in cattle, swine and horses are *L. pomona*, *L. grippotyphosa*, *L. canicola*, and *L. icterohemorrhagiae*. *L. hardjo* has also been isolated from cattle and has increased in prevalence in recent years (Figure 25-3).

**25-1.2b Methods of Transmission**   The *Leptospira* organism may be ingested thereby entering the circulatory system, but tends to congregate in extremely large numbers in the kidneys. Organisms from the kidneys are

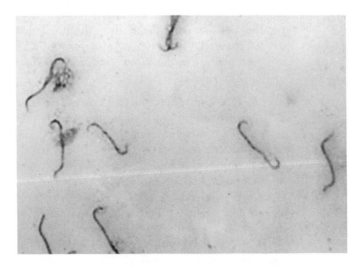

FIGURE 25-3. Photomicrograph of *Leptospira* organisms. (Courtesy of R. Hidalgo. College of Vet. Med., Miss. State Univ.)

eliminated in the urine, which is the principal source of the organism for transmission. Cattle recovering from the disease continue to shed the organism from the kidneys for 2 to 3 months or longer. There is evidence that swine and some species of wild animals may become permanent spreaders of the organism. Transmission is accomplished by the infected animal contaminating pastures, water, and other feed supply with urine. In management systems where animals are concentrated on concrete feedlots, the air becomes contaminated with small water particles containing many organisms. The organisms are thus ingested or inhaled by the susceptible animal.

It is possible for the *Leptospira* organisms to be transmitted through semen. Infected bulls may transmit the disease to cows during natural mating. Commerical AI organizations do not collect and process semen from bulls shedding *Leptospira* organisms in their semen.

**25-1.2c Symptoms**   Several symptoms have been described for leptospirosis in cattle. Depending on the severity of the disease, some or all of the symptoms may be observed in the same animal. The incubation period is usually 7 to 9 days. However, longer periods are possible. Cattlemen should watch for:

1. *Sudden onset with loss of appetite.* In acute cases animals lose interest in water and feed.
2. *Loss of body weight.* In acute cases in mature dairy cows body

weight losses up to 136 kg in 3 days have been reported. Some of this weight loss is due to shrinkage as a result of the animal not eating or drinking. However, much of it is true weight loss.

3. *Drop in milk production.* The effect on milk production will vary with the severity of the disease. In severe cases in high producing dairy cows it may drop almost to zero. The little milk obtained from these cows usually is abnormal, thick and dark yellow in color or it may contain blood. Because of the abnormal udder secretion it has been suggested that the *Leptospira* organism causes mastitis. The abnormal secretion appears to be a result of abnormal body metabolism rather than an invasion of the mammary gland by the organism.

4. *Elevated body temperature.* Anaimals of various ages may show elevated body temperatures as high as 40.5 to 41.7°C.

5. *Blood in the urine.* The urine of some animals will be tinted red. The blood cells are hemolized so that the red color is released from the cells and does not settle on standing. The blood in the urine is a result of damage to the kidneys by the organisms.

6. *Abortions.* The incidence of abortion in a herd infected with leptospirosis may vary from a few to as high as 40%. The incidence depends on the stage of pregnancy of the herd as a whole. Most lepto-initiated abortions occur during the latter third of pregnancy, 7 to 10 days after the onset of the disease. It was once thought that the high body temperature resulted in the death of the fetus. However, the organisms have the ability to cross the placental barrier from the maternal circulation to invade the fetus. Presumably this passage is made easier during advanced stages of pregnancy when some hemorrhage occurs at the hylus of the placentome. This concept is supported by the fact that aborted feti are extensively autolyzed (enzymatic degeneration of cells). In some herd outbreaks, abortion without apparent illness may be the only sign of the disease.

7. *Reduced breeding efficiency.* There is little experimental evidence to indicate that cows recovering from leptospirosis continue to experience reduced breeding efficiency. However, many astute managers report that reproductive problems do continue up to one year in the form of increased services per conception and retained placentae. These problems seem to be more pronounced with the serotype *L. hardjo* than with the others. Animals recovering from severe leptospirosis frequently have a slow and prolonged convalescence.

In swine leptospirosis not only causes abortions but often infects pigs that are carried to term so that they are weak at birth and die in a few days. Mummified feti may also be seen along with weak pigs. The disease has been reported to result in abortion in mares. The clinical disease and abortion occur less frequently in horses than in cattle and swine.

**25-1.2d Diagnostic procedures** The most common diagnostic procedure used for leptospirosis is the serum agglutination test utilizing the tube or plate procedures. A positive titer does not necessarily mean that an animal is currently infected with leptospirosis. It could mean that the animal has previously had the disease or that it had been vaccinated. There is value in repeating the tests in 2 or 3 weeks and comparing the two. Animals negative to the first test and positive to the second test or rising titers in animals that were positive on the first test would indicate active disease. Testing for all serotypes is recommended when leptospirosis is suspected.

Isolation and identification of the organism is a more positive diagnostic procedure. However, there are some problems associated with culturing *Leptospira* organisms in that *Leptospira* do not grow well on most artificial culture media. The samples, whether urine or blood, must be taken at a time when the organisms are present. These organisms are very delicate and die readily outside of the host.

**25-1.2e Control measures** There is very little that can be done to prevent animals from being exposed to *Leptospira* organisms. The wide variety of susceptible animals in both wild and domesticated species makes the exposure to *Leptospira* organisms a continous threat. Avoiding access to farm ponds and slow moving streams may be helpful. Isolation and vaccination of newly purchased animals are recommended.

Vaccination of cattle, horses, and swine annually is indicated. Multivalent bacterins are available for cattle containing the five serotypes described previously. Some early *L. pomona* bacterins contained some foreign proteins which caused anaphylactic (allergic) reactions when used repeatedly. This is no longer a problem. For management reasons swine may be vaccinated before each breeding period.

**25-1.3 Brucellosis**

Brucellosis is a worldwide problem with both public health and economic importance. The disease may be caused by three major and one minor species of the genus *Brucella*. Each is specific for a species of farm animals but all are contagious to man.

Most human cases of brucellosis are contracted from exposure in slaughter houses, with infection largely traced to sows. Some cases have been traced to cattle both in the slaughter house and on farms. At one time most human brucellosis was due to the consumption of raw milk. This is no longer true since very little raw milk is consumed. More important, however, brucellosis has essentially been eliminated from dairy cattle.

Young animals apparently carry a rather strong passive immunity to the *Brucella* organism. Offspring from infected females usually are infected at the time of birth and they consume many organisms while nursing their dams. These organisms apparently do not cause any ill effects and usually are eliminated within a few weeks. If an adult nonpregnant cow becomes infected the organisms localize in the mammary gland and later spread to the uterus during pregnancy. The organisms invade the pregnant uterus, placenta, and fetus.

**25-1.3a Causative organisms**  In cattle *B. abortus* is the organism responsible for the disease. *B. melitensis* is the major species causing brucellosis in sheep but *B. ovis* also causes a problem especially in rams. *B. suis* is responsible for swine brucellosis.

**25-1.3b Symptoms**  The most prominent clinical sign of brucellosis in cattle is abortion after the fifth month of gestation. Subsequent pregnancies may result in a normal calf carried to term but two or three abortions may occur in the same animal. Some full term calves are weak and may not survive. Retained placentae and uterine infections are both common after effects of brucellosis. In severe cases of uterine infection sterility may occur. The testes of the bull may become infected and this orchitis can result in sterility.

*Brucella melitensis* causes abortions in the ewe and probably other symptoms similar to those in cattle. *B. ovis* infections are more serious in the ram causing epididymitis which causes lesions in the epididymis and if the lesions are bilateral sterility may result. Ewes may abort late in gestation but this is not common.

The most common symptom of brucellosis in swine is abortion and the birth of weak pigs. Sows may have a slight uterine discharge indicating a uterine infection that persists for an extended period of time. Such uterine infections result in temporary, if not permanent, sterility. Sows that abort following the onset of the disease usually carry litters to term thereafter. The testes of the male may become infected and result in sterility.

**25-1.3c Methods of transmission**  The usual method of transmission in livestock is by ingesting contaminated feed or water. The aborting female sheds large numbers of *Brucella* organisms along with the aborted fetus and fluids associated with it. Females that carry their offspring to term have a build-up of organisms in the reproductive tract and the fluids and discharges following parturition contaminate the environment. While the disease may be transmitted at other times, the most contagious period certainly is during and immediately following abortion or parturition.

The disease can also be transmitted through the semen from infected males. Bulls used for semen production in AI are frequently tested for brucellosis to be sure that no semen is processed from bulls infected with brucellosis.

**25-1.3d Diagnostic procedures**  The most common test procedure for brucellosis is the agglutination test utilizing blood serum. The plate, tube, and card tests are variations which are used. The card test is the most sensitive and has been used as a field test. Because of its level of sensitivity, one of the other tests should be utilized to confirm positive card tests. Until recently it has not been possible to differentiate between positive titers caused by vaccination and those caused by the field strain of Brucella organism. The Rivanol and complement fixation tests have been adapted to supplement the serum agglutination test and with these, field strain antibody titers can be differentiated from vaccination antibody titers.

The brucellosis ring test (BRT) is used as a screening test for dairy herds. A bulk milk sample from the entire herd can be tested for the presence of antibodies. The test is sensitive enough to detect the presence of one infected cow in a herd of 200 or more cows. When a positive BRT occurs, each individual cow in the herd must be blood tested to determine which ones are infected.

**25-1.3e Control measures**  A nonpathogenic strain of *Brucella abortus* known as strain 19 has been used to produce a live organism vaccine to evoke immunity in cattle. The vaccine has been used exclusively as a calfhood vaccination (prior to puberty). The recommended vaccination age is 3 to 6 months for dairy cattle and 3 to 8 months for beef cattle. The efficacy of the vaccine is not sufficient to provide 100% protection in high exposure situations, however, it is effective as a preventive measure. With the introduction of the Rivanol and complement fixation tests strain 19 vaccine is being utlized as an adult vaccine in many problem herds in high-incidence areas. Bulls are not vaccinated with strain 19 since the organisms localize in the testes and cause sterility in some cases.

There has been a state-federal cooperative eradication program in effect since 1955. Progress has been slow, but much of the United States is virtually free of brucellosis. A two-year study by a Brucellosis Advisory Council has declared that the goal of total eradication of brucellosis can be accomplished and should be pursued vigorously.

Test-and-slaughter has been utilized as the primary control measure for bovine brucellosis. Calfhood vaccination has been used as a supplemental control measure and in many instances has not been used extensively.

Vaccination plus test-and-slaughter are used to control brucellosis in sheep. Rams with epididymitis should be slaughtered. Immature rams may be vaccinated with two doses of a killed *B. ovis* vaccine.

The control of brucellosis in swine is based on test-and-slaughter. In many cases an entire herd may need to be slaughtered. There is no commerical vaccine available to evoke immunity in swine.

**25-1.4
Listeriosis**

This disease occurs in many species of mammals and birds but is not especially common in domestic livestock. It has significance in that it causes similar disease symptoms in cattle, sheep, and man. It is caused by the bacterium *Listeria monocytogenes* and commonly invades the brain and meninges (membranes covering the brain).

In addition to its effect on the central nervous system, a low incidence of abortions occurs in cattle. The abortions usually occur between the fourth and seventh month of gestation. Encephalitis and abortion may occur in the same animal, but outbreaks in herds usually result in either encephalitis or abortions. The abortion rate in sheep may be more common than in cattle with reported rates ranging to 50%. The abortions in sheep are usually near term.

One method of transmission is the ingestion of the organism with contaminated feed or water. Isolation of the organism by culture is the best means for diagnosis. Penicillin and tetracycline antibiotics have been used as therapy to reduce mortality rate. The public health aspect of the disease makes sanitation, especially at the time of abortions, essential. Human cases may follow exposure resulting from handling aborted feti.

**25-1.5 Nonspecific uterine infections**

Many studies have been conducted on the microorganisms in the uterus of repeat breeder cows compared to normal cows in an attempt to determine the cause of reduced fertility. These studies have revealed very little since most uteri contain organisms capable of causing infections. Inoculations of the uterus with organisms capable of pathogenicity have shown little effect on breeding efficiency. It appears that the uterus has a good defense mechanism and becomes infected only when this mechanism is interrupted. It is generally agreed that the uterine defense is much greater during estrus than it is during the luteal phase of the estrous cycle.

Most serious uterine infections occur at the time of, or immediately following, parturition. Many are associated with retained placentae. *Metritis* is described as an infection of the endometrium of the uterus and is characterized by the elimination of pus. The endometrium thickens and the myometrium loses its tone. The total uterine wall is spongy and unresponsive.

*Endometritis* is an inflammatory condition of the uterine mucosa with the absence of pus. The characteristics of the uterine wall are similar to those for metritis.

No particular pattern as to causative organisms has been established. *Streptococci, Staphylococci, Diplococci, Escherichia coli* are examples of isolations from the uteri of cows with metritis and endometritis. Control measures involve sanitation during parturition, proper treatment of retained placentae (see Chapter 19), and treatment with broad spectrum antibiotics.

**25-1.6 Contagious equine metritis**

Contagious equine metritis (CEM) is a highly contagious acute venereal disease of horses. The disease was first officially reported in the United Kingdom during the summer of 1977. Later reports reveal that CEM occurred on at least four stud farms in Ireland in 1976. Unconfirmed reports indicate that CEM may have occurred in France before 1976. Contagious equine metritis was diagnosed in Kentucky in the spring of 1978 and in Missouri in 1979, but has been confined to a few stud farms. The disease has been reported only in thoroughbred horses.

The causative organism is a bacterium but has not been classified. It has characteristics of the genus *Moraxella* and some similarity to *Haemophilus*. Transmission has been primarily, if not entirely, by natural mating. AI has not been allowed in the thoroughbred ranks. It is conceivable that the organism may be transferred by humans using contaminated equipment.

Symptoms of the disease are metritis associated with cervicitis and vaginitis. A copious mucopurulent discharge from the vulva may occur. The severity varies from mares that have an almost complete sloughing of the endometrium to mares that appear almost clinically normal but fail to conceive. All infected mares are at least temporarily sterile.

The stallion shows no clinical symptoms of the disease yet remains a carrier and transmits the disease for an extended period of time. The stallion with CEM appears to play the same role in the spread of the disease that the bull does in spreading trichomoniasis.

Diagnosis is presumptive when several mares bred to a particular stallion come back into estrus and show other clinical symptoms. Cervical and clitoral swabs may be taken for bacteriological culture. Swabs should be taken from the penile sheath and the urethral fossa and the urethra of the stallion for culture. There is no serological test available to date.

Control measures include quarantine of infected animals. The United States and Canada have placed a ban on importation of all equidae except geldings, weanlings, and yearlings from Ireland and the United Kingdom, France, and Australia. Strict sanitation should be practiced when working with the breeding herd.

Treatment of infected mares involves several days of systematic treatment with large doses of antibiotics along with intrauterine infusions. Less intense therapy appears to induce a carrier state in some mares. Treating the stallion includes cleansing the penis and prepuce thoroughly with chlorhexidine and topical treatment with nitrofurazone daily for at least 3 days or with three treatments over a 5-day period. In addition, intramuscular treatment with antibiotics for 5 to 10 days is desirable. Further treatment may be necessary depending on the results of bacteriological tests or trial breedings. Economics must be considered in the treatment of mares and stallions.

# 25-2
# Protozoan
# diseases

## 25-2.1
## Bovine
## trichomoniasis

Trichomoniasis is a venereal disease of cattle caused by a protozoan *Trichomonas fetus*. The organism is 10 to 25$\mu$ in length and has 3 anterior flagella and an undulating membrane with one posterior flagellum (Figure 25-4).

FIGURE 25-4.   *Trichomonas fetus* protozoan. (From Asdell and Bearden. 1959. *Cornell Ext. Bull.* 737.)

**25-2.1a Methods of transmission**   Since trichomoniasis is a venereal disease it is mainly transmitted by sexual contact. Sexual contact in this case includes artificial insemination. There is no satisfactory way to treat semen from infected bulls to make it safe for use in AI.

**25-2.1b Symptoms**   The symptoms of trichomoniasis are similar to those of *Vibrio fetus* infection. Many cows conceive, but the organism kills the embryo and causes the cow to return to estrus at irregular, long intervals. A major difference between the two diseases is the development of pyometra in trichomoniasis, characterized by the accumulation of pus along with the degeneration of a fetus in the uterus. The liquefaction necrosis of the fetus in the pus may continue for an extended period of time. Herds infected with trichomoniasis will have a high incidence of discharge containing pus from the reproductive tract.

**25-2.1c Diagnosis**   Diagnosis is made by direct microscopic examination of exudates from the female or prepucial washings from the male. It is a tedious process and negative results should not be considered proof of freedom from infection. Further testing may be needed.

**25-2.1d Control measures**   When the disease is diagnosed in a herd a period of sexual rest for the females is the only treatment. A rectal examination may be performed to determine those having pyometra. These animals should be selectively treated with antibiotics. The cervical plug may be broken to allow the exudate to leave the uterus. Approximately 90 days will be required for the cows to recover and to eliminate the organisms from their reproductive tracts. In most cases bulls should be slaughtered. Success in eliminating the organisms from bulls with topical applications of sodium iodide, acroflavin, and bovoflavin ointment has been reported. The time and expense of the treatment and the testing procedures to determine if the bull is free of the organism can only be justified in the very valuable bulls. Sexual rest for cows followed by at least one year of artificial insemination is recommended. No vaccine is commercially available.

**25-2.2
Toxoplasmosis**

This organism causes disease in sheep, cattle, swine, and man. The causative organism is *Toxoplasma gondii*, a small elongated organism frequently observed in compact masses. Transmission is by ingestion of sporulated oocysts either from cat feces or flesh of infected animals.

The symptoms associated with reproduction are abortion, premature births, and stillbirths. Some full-term offspring appear to be infected as a result of intrauterine transmission resulting in high mortality. Abortions in sheep occur during the last 4 to 6 weeks of gestation.

Diagnosis is based on the isolation of the organism or by serological tests. Microscopic examination of necrotic areas of the placentome, or impression smears of the necrotic area, may reveal the organism. The Sabin-Feldman dye test is a valuable serological test for the organism.

Control measures depend on breaking the infection cycle. Further studies concerning the role of the cat in transmission may be helpful. No vaccine is commercially available.

**25-3
Viral diseases**

**25-3.1
Bovine viral
diarrhea (BVD)**

The bovine viral diarrhea virus affects a high percentage of the cattle in the United States. In many herds where no clinical symptoms of the disease have been observed, as high as 50% of tested cattle exhibit antibodies against the virus. Four clinical forms of the disease have been described: (1) A subclinical form with which no symptoms are observed is perhaps the most common. (2) Chronic viral diarrhea may occur insidiously and not be recognized for some time. Loss of appetite, emaciation, mild diarrhea, and subnormal growth rate are symptoms that accompany this form. (3) Acute viral diarrhea is characterized by profuse diarrhea, elevated body temperature, and erosion of the gastrointestinal tract. Herd outbreaks have been reported in which essentially 100% of the animals were infected and in other cases where only a few individuals were affected. In the latter case, it is probable that the remainder of the animals were immune from previous infection. (4) The mucosal disease form is the most severe and is characterized by all of the symptoms described for the acute form plus an ulceration of the oral cavity and the mucous membranes of the digestive tract. The animals die in about 14 days without developing immunity. This form of the disease is seen more frequently in animals from 8 to 18 months of age.

In pregnant cows the virus may invade the feti causing death and abortion. The dead fetus is usually held in the uterus for several days before expulsion. Most abortions caused by the BVD virus occur during the first 3 to 4 months of gestation. The incidence of abortion caused by BVD is much lower than early data suggested. Concurrent diseases or the appearance of antibodies against BVD in serum following abortions has been taken as circumstantial evidence that BVD was the cause of the abortions.

BVD infections during the middle third of gestation have resulted in developmental defects of the brain, eye, and hair in calves. Brain and eye defects occur more frequently than hair defects.

The disease is transmitted by the viral agent coming in contact with mucous membranes from a contaminated environment. The virus may be transmitted in the semen from infected bulls either through AI or natural services. It is diagnosed by isolation of the virus or by fluorescent antibody technique, serum neutralization test, and tissue culture procedures. The virus is difficult to isolate from aborted feti because of the state of deterioration by the time of expulsion.

Control measures depend entirely on vaccination with a modified live virus vaccine. The vaccination of replacement animals between 8 and 12 months of age is a common practice. Immunity may last for the life of the animal, but in closed herd situations where little challenge is provided annual booster vaccinations may be advisable. Pregnant animals should not be vaccinated. In herds where the disease has been diagnosed, calves 2 months of age and older should be vaccinated. Any animal vaccinated under 8 months of age should be revaccinated at about 12 months of age. These younger animals usually have some passive immunity from their dams which interferes with the establishment of lasting immunity.

**25-3.2 Infectious bovine rhinotracheitis-infectious pustular vulvovaginitis (IBR-IPV)**

The IBR-IPV virus, like others, is transmitted by viral particles from contaminated environment coming in contact with mucous membranes. The symptoms produced depend on the systems affected. As a respiratory ailment it causes coughing, wheezing, nasal discharge, and fever. As a digestive system ailment, particularly in calves, it causes diarrhea. When the organism invades the reproductive system a profuse pustular vulvovaginitis results. A discharge containing pus may be observed from the vulva. Temporary infertility is experienced and it is usually a good idea to suspend breeding during the early phases of the disease. The glans penis of the bull usually develops pustules and the associated pain may be sufficient to prevent breeding.

Cows with either respiratory or vulvovaginitis forms of the disease may experience viral invasion of the fetus resulting in abortion. The feti are generally aborted from 3 weeks to 3 months after infection of the dam. Retained placentae are commonly associated with the disease.

The disease is diagnosed by isolation of the virus by tissue culture or by serum neutralization tests. Control measures involve vaccination with modified live virus vaccine as a preventive measure. Both intramuscular and nasal spray preparations of the vaccine are available. Replacement animals after 6 months of age and nonpregnant adult females may be vaccinated with the IM preparation. This vaccination may give lifetime immunity but booster vaccinations with the nasal spray preparation may be advisable. Most nasal vaccines may be used on pregnant cows without fear of abortion.

**25-3.3
Equine viral
rhinopneumonitis**

Equine rhinopneumonitis virus is also known as equine herpes virus I. It causes a respiratory problem in young horses and is a major cause of abortion. Symptoms include nasal discharge, fever, depression, and coughing. The duration of the disease is usually about 4 weeks followed by spontaneous recovery. The disease is transmitted by viral particles coming in contact with mucous membranes primarily from aerosols. Immunity following the disease is usually transient and the animals are susceptible again after several months.

The nursing foals that are infected provide massive exposure to the mares which are usually in midpregnancy. The rate of abortion in mares depends on their susceptibility and has been reported as high as 80% in some outbreaks. The fetus dies *in utero* and is promptly expelled without evidence of autolysis. Most abortions occur from the eighth month to term. Mares usually show no symptoms of the illness, the placentae are not retained, and fertility is normal following abortion.

Diagnosis of the disease is based on viral isolation and identification and from gross and microscopic lesions in the aborted fetus.

A control program based on planned infection to produce immunity has been described. A hamster adapted virus has been used to expose all horses on a premise twice yearly, and in most management programs this is in July and October. The exposure should be timed so that mares are not vaccinated after the fifth month of pregnancy. Even though the merits of this control program have been demonstrated, a safer and more effective vaccine is needed.

**25-3.4
Equine viral
arteritis**

A disease that was very common in horses during the early part of the century was probably caused by equine arteritis virus. The disease presently occurs only sporadically in the United States. As the name implies, the disease causes an inflammatory reaction in the small arteries. Pulmonary edema and edema of the limbs are common. Other symptoms may be elevated body temperature, severe depression, loss of appetite, weight loss, colic, and diarrhea. A brick red mucous membrane protrudes around the eye with watery and sometimes purulent discharge.

Abortions occur in 50 to 80% of the infected mares. Abortions tend to occur in the latter half of gestation. The tissues and fluids associated with the infected aborted feti contain large numbers of viral particles. Transmission is primarily by the viral particles coming in contact with the mucous membranes due to inhaling aerosols containing these particles.

Diagnosis is based on clinical symptoms and the elimination of other causes of abortion. Control measures include isolation of apparently infected mares particularly at the time of abortion and parturition because of the large numbers of particles elminated with the fluids at the time of abortion or parturition. A modified live virus vaccine has been prepared but its commercial availability will depend on the prevalance of the disease and the need for immunization.

**25-3.5**
**Bluetongue**

The virus that causes the disease Bluetongue in sheep and cattle is transmitted by anthropod-vector. Until recently the disease has been considered primarily a problem of the western states. A recent survey in which blood samples from all 48 contiguous states were shipped to the National Animal Disease Laboratory at Ames, Iowa and tested for antibodies against the Bluetongue virus revealed cattle with antibodies in essentially all states. One area in the central Atlantic states (Georgia, North and South Carolina) and another near midcontinent (Nebraska to Texas) approached the epizootic stage.

General symptoms of the disease in sheep are elevated temperature, severe depression, and loss of appetite. Erosions occur on the lips, cheeks, and tongue with the latter appearing cyanotic (turns blue due to $O_2$ deficiency) hence the name of the disease.

Reproductive effects are primarily on the fetus. Although lambs are born at full term some are stillborn, others are spastic and lie struggling until death. Some are referred to as "dummy lambs" and are unable to stand or may be uncoordinated and fail to nurse. The most severe effects occur when the disease is contracted while the ewes are in the fourth to eighth week of gestation. Infections at later stages may produce problems but not so severe.

Cattle have been suggested as the principal reservoir for the virus. Clinically in cattle the disease affects the mouth and feet, causes abortion and weak or stillborn calves. Loss of muscular coordination and blindness occur. Healthy carriers are an important threat for transmission. The virus is found in bull semen and probably can be transmitted by natural mating and through AI.

Control measures involve vaccination with a modified live virus vaccine which provides good immunity for a period of 2 to 4 years. Pregnant ewes cannot be vaccinated because the vaccine produces conditions similar to the disease itself. Ewes should be vaccinated at least 4 weeks prior to the breeding season.

**25-4**
**Mycoses**

Several mycotic (fungus) infections have been associated with abortion in cattle. The principal organism is *Aspergillus fumigatus*. *A. absidia* and *A. mucor* have also been isolated from aborted feti. Abortions usually occur in middle to late gestation. Transmission is by inhalation of spores from moldy feed. Both concentrates and forages improperly stored may become moldy and contribute to the problem. Diagnosis may be based on demonstrating mycelium in the fetal membranes or by culturing the fungus from stomach content of aborted feti. Control measures are best accomplished by proper feed handling and preservation to prevent mold.

**Suggested reading**

Bartlett, D. E., K. Moist and F. A. Spurrell. 1953. The *Trichomonas fetus*-infected bull in artificial insemination. *J. Amer. Vet. Med. Assoc.*, 122:366.

Doll, E. R. and J. T. Bryans. 1963. A planned infection program for immunizing mares against viral rhinopneumonitis. *Cornell Vet.*, 53:249.

Gibbons, W. J., E. J. Catcott and J. F. Smithcors. 1970. *Bovine Medicine and Surgery*. Amer. Vet. Publications, Inc.

Jones, T. C. 1969. Clinical and pathological features of equine viral arteritis. *J. Amer. Vet. Med. Assoc.*, 155:315.

Kahrs, R. F., F. W. Scott and A. deLahunta. 1970. Bovine viral diarrhea-mucosal disease, abortion and congenital cerebella hypoplasia in a dairy herd. *J. Amer. Vet. Med. Assoc.*, 156:851.

Kendrick, J. W. and J. A. Horwarth. 1974. Reproductive diseases. *Reproduction in Farm Animals*. (3rd ed.) ed. E. S. E. Hafez. Lea and Febiger.

Kendrick, J. W. and O. C. Straub. 1967. Infectious bovine rhinotracheitis-infectious pustular vulvovaginitis virus infection in pregnant cows. *Amer. J. Vet. Res.*, 20:1269.

McEntee, K., D. W. Hughes and W. C. Wagner. 1959. Failure to produce vibriosis in cattle by vulvar exposure. *Cornell Vet.*, 49:34.

Olds, D. 1978. Inherited, anatomical, and pathological causes of lowered reproductive efficiency. *Physiology of Reproduction and Artificial Insemination of Cattle*. (2nd ed.) ed. G. W. Salisbury, N. L. VanDemark and J. R. Lodge. W. H. Freeman and Co.

Osebold, J. W. 1977. Infectious diseases influencing reproduction. *Reproduction in Domestic Animals*. (3rd ed.) ed. H. H. Cole and P. T. Cupps. Academic Press.

Walker, J. S. 1978. *Contagious Equine Metritis Reference Handbook*. Plum Island Anim. Dis. Ctr., USDA, SEA, FR.

# Index

## DATE DUE

|  |  |  |  |
|--|--|--|--|
|  |  |  |  |
|  |  |  |  |
|  |  |  |  |
|  |  |  |  |
|  |  |  |  |
|  |  |  |  |
|  |  |  |  |
|  |  |  |  |
|  |  |  |  |
|  |  |  |  |
|  |  |  |  |
|  |  |  |  |
|  |  |  |  |